T0189906

NEW BIOTECHNOLOGIES FOR INCREASED ENERGY SECURITY

The Future of Fuel

NEW BIOTECHNOLOGIES FOR INCREASED ENERGY SECURITY

The Future of Fuel

Edited by
Juan Carlos Serrano-Ruiz, PhD

Apple Academic Press Inc.	Apple Academic Press Inc.
3333 Mistwell Crescent	9 Spinnaker Way
Oakville, ON L6L 0A2	Waretown, NJ 08758
Canada	USA

©2016 by Apple Academic Press, Inc.

First issued in paperback 2021

Exclusive worldwide distribution by CRC Press, a member of Taylor & Francis Group

No claim to original U.S. Government works

ISBN 13: 978-1-77463-560-5 (pbk)
ISBN 13: 978-1-77188-146-3 (hbk)

Library and Archives Canada Cataloguing in Publication

New biotechnologies for increased energy security : the future of fuel/edited by Juan Carlos Serrano-Ruiz, PhD.

Includes bibliographical references and index.
ISBN 978-1-77188-146-3 (bound)
1. Biomass energy. 2. Biotechnology. 3. Energy security. I. Serrano-Ruiz, Juan Carlos, author, editor

TP339.N43 2015	662'.88	C2015-901591-X

Library of Congress Cataloging-in-Publication Data

New biotechnologies for increased energy security : the future of fuel / editor, Juan Carlos Serrano-Ruiz, PhD.

pages cm
Includes bibliographical references and index.
ISBN 978-1-77188-146-3 (acid-free paper)
1. Biomass energy. 2. Biotechnology. I. Serrano-Ruiz, Juan Carlos.

TP339.N484 2015	662'.88--dc23	2015007772

Apple Academic Press also publishes its books in a variety of electronic formats. Some content that appears in print may not be available in electronic format. For information about Apple Academic Press products, visit our website at **www.appleacademicpress.com** and the CRC Press website at **www.crc-press.com**

ABOUT THE EDITOR

JUAN CARLOS SERRANO-RUIZ

Juan Carlos Serrano-Ruiz studied Chemistry at the University of Granada (Spain). In 2001 he moved to the University of Alicante (Spain) where he received a PhD in Chemistry and Materials Science in 2006. In January 2008, he was awarded a MEC/Fulbright Fellowship to conduct studies on catalytic conversion of biomass in James Dumesic's research group at the University of Wisconsin-Madison (USA). He is (co)author of over 50 manuscripts and book chapters on biomass conversion and catalysis. He is currently Senior Researcher at Abengoa Research, the research and development division of the Spanish company, Abengoa (Seville, Spain).

CONTENTS

Part IV: Genetic Engineering

Part V: Nanotechnology and Chemical Engineering

ACKNOWLEDGMENT AND HOW TO CITE

The editor and publisher thank each of the authors who contributed to this book. The chapters in this book were previously published elsewhere. To cite the work contained in this book and to view the individual permissions, please refer to the citation at the beginning of each chapter. Each chapter was read individually and carefully selected by the editor; the result is a book that provides a nuanced look at the intersection between developing biotechnologies and the future of our energy security. The chapters included examine the following topics:

- The editorial found in chapter 1 is a good introduction to the urgent relevancy of this topic.
- In chapter 2, Rowley and his colleagues determine methodology for improving the efficiency of extracting xylan from corn stover using dimethyl sulfoxide combined with heat, significant because of third-generation bioenergy's focus on non-food biomass stock.
- Chapter 3 contains an investigation by Rana et al. of pretreatment methods and enzymatic hydrolysis for producing higher glucose yields from corn stover as a biomass for bioenergy conversion.
- In chapter 4, we have research that supports acid-catalyzed thermochemical pretreatment of lignocellulosic feedstocks as a simple and inexpensive approach for pretreatment that efficiently improves the susceptibility to cellulolytic enzymes, even for more recalcitrant types of lignocellulose.
- Zhang and colleagues found in chapter 5 that sabinene was significantly produced by assembling a biosynthetic pathway and evaluated other methodologies for optimizing sabinene production.
- In chapter 6, my colleagues and I investigated technologies that indicate that advanced biofuels such as green hydrocarbons represent an attractive alternative to conventional bioethanol and biodiesel.
- Because isoprenoids are an excellent illustration of the chemical diversity and unique biochemical roles that are possible within members of a single molecular family, in chapter 7, Jarchow-Choy and her colleagues investigate these structures and their roles, particularly in terms of synthetic methodologies and enzymological studies.

- Yang and Liu MPE discuss in chapter 8 metabolic process engineering's role in an efficient fermentation process for biochemical and biofuel production.
- In chapter 9, Rasala and her colleagues report the construction and validation of a set of transformation vectors that enable protein targeting to distinct subcellular locations; they then present two complementary methods for multigene engineering in the eukaryotic green microalga *C. reinhardtii*, a viable option for biofuel production.
- Because *R. eutropha* has great potential to directly produce biofuels, Bi and colleagues demonstrate in chapter 10 the engineering utility of a plasmid-based toolbox for *R. eutropha*.
- In chapter 11, Colmenares and Luque provide an overview of recent investigations into selective photochemical transformations using nanomaterials, particularly focused on photocatalysis for lignocellulose-based biomass valorization as an important option for sustainable energy production.
- Das and El-Safty investigate in chapter 12 integrating sulfate into the development of zirconium nanoparticles, concluding that this offers an excellent heterogeneous biodiesel catalyst for the effective conversion of long-chain fatty acids to their methyl esters, a process vital for the production of certain biofuels.
- Chapter 13 provides us with the investigation of Feyzi and colleagues into sunflower oil transesterification with methanol.

LIST OF CONTRIBUTORS

Birgitte K. Ahring
Bioproducts, Sciences and Engineering Laboratory (BSEL), Washington State University, 2710 Crimson Way, Richland, WA 99354-1671, USA

Björn Alriksson
Processum Biorefinery Initiative AB, Örnsköldsvik, SE-891 22, Sweden

Daniel J. Barrera
California Center for Algae Biotechnology and Division of Biological Sciences, University of California San Diego, La Jolla, California, United States of America

Harry R. Beller
Earth Sciences Division, Lawrence Berkeley National Laboratory, Berkeley, CA 94720, USA

Changhao Bi
Physical Biosciences Division, Lawrence Berkeley National Laboratory, Berkeley, CA 94720, USA; Present address: Tianjin Institute of Biotechnology, Chinese Academy of Sciences, Tianjin, China

Stuart Black
National Renewable Energy Laboratory, Golden, CO, USA

Yujin Cao
CAS Key Laboratory of Biobased Materials, Qingdao Institute of Bioenergy and Bioprocess Technology, Chinese Academy of Sciences, No.189 Songling Road, Qingdao, Laoshan District 266101, China

Syh-Shiuan Chao
California Center for Algae Biotechnology and Division of Biological Sciences, University of California San Diego, La Jolla, California, United States of America

Swapnil R. Chhabra
Physical Biosciences Division, Lawrence Berkeley National Laboratory, Berkeley, CA 94720, USA

Juan Carlos Colmenares
Institute of Physical Chemistry, Polish Academy of Sciences, ul. Kasprzaka 44/52, 01-224 Warsaw, Poland.

Swapan K. Das
National Institute for Materials Science (NIMS), 1-2-1 Sengen, Tsukuba, Ibaraki 305-0047 (Japan), Fax: (+81)29-859-2501

Stephen R. Decker
National Renewable Energy Laboratory, Golden, CO, USA

Sherif A. El-Safty
National Institute for Materials Science (NIMS), 1-2-1 Sengen, Tsukuba, Ibaraki 305-0047 (Japan), Fax: (+81)29-859-2501; Graduate School for Advanced Science and Engineering, Waseda University, 3-4-1 Okubo, Shinjuku-ku, Tokyo 169-8555 (Japan)

Xinjun Feng
CAS Key Laboratory of Biobased Materials, Qingdao Institute of Bioenergy and Bioprocess Technology, Chinese Academy of Sciences, No.189 Songling Road, Qingdao, Laoshan District 266101, China

Mostafa Feyzi
Faculty of Chemistry, Razi University, P.O. Box 6714967346, Kermanshah, Iran; Nanoscience & Nanotechnology Research Center (NNRC), Razi University, P.O. Box 6714967346, Kermanshah, Iran

David T. Fox
Los Alamos National Laboratory, USA

Nathan J. Hillson
Physical Biosciences Division, Lawrence Berkeley National Laboratory, Berkeley, CA 94720, USA

Sarah K. Jarchow-Choy
Los Alamos National Laboratory, USA

Leif J. Jönsson
Department of Chemistry, Umeå University, Umeå SE-901 87, Sweden

Andrew T. Koppisch
Northern Arizona University, USA

Hui Liu
CAS Key Laboratory of Biobased Materials, Qingdao Institute of Bioenergy and Bioprocess Technology, Chinese Academy of Sciences, No.189 Songling Road, Qingdao, Laoshan District 266101, China

Qiang Liu
College of Food Science, Sichuan Agricultural University, Yaan 625014, China

Xiaoguang Liu
Department of Chemical and Biological Engineering, The University of Alabama, USA

Rafael Luque
Departamento de Quimica Organica, Universidad de Cordoba, Campus de Rabanales, Edificio Marie Curie, E-14014, Cordoba, Spain

Stephen P. Mayfield
California Center for Algae Biotechnology and Division of Biological Sciences, University of California San Diego, La Jolla, California, United States of America

William Michener
National Renewable Energy Laboratory, Golden, CO, USA

Jana Müller
Physical Biosciences Division, Lawrence Berkeley National Laboratory, Berkeley, CA 94720, USA

Nils-Olof Nilvebrant
Borregaard, Sarpsborg, 1701, Norway

Leila Norouzi
Faculty of Chemistry, Razi University, P.O. Box 6714967346, Kermanshah, Iran

Matthew Pier
California Center for Algae Biotechnology and Division of Biological Sciences, University of California San Diego, La Jolla, California, United States of America

Pattanathu K. S. M. Rahman
School of Science & Engineering, Technology Futures Institute, Teesside University, Middlesbrough, UK

Hamid Reza Rafiee
Faculty of Chemistry, Razi University, P.O. Box 6714967346, Kermanshah, Iran

Diwakar Rana
Bioproducts, Sciences and Engineering Laboratory (BSEL), Washington State University, 2710 Crimson Way, Richland, WA 99354-1671, USA

Vandana Rana
Bioproducts, Sciences and Engineering Laboratory (BSEL), Washington State University, 2710 Crimson Way, Richland, WA 99354-1671, USA

Beth A. Rasala
California Center for Algae Biotechnology and Division of Biological Sciences, University of California San Diego, La Jolla, California, United States of America

John Rowley
University of Colorado, Boulder, CO, USA

Kamaljeet Kaur Sekhon
School of Science & Engineering, Technology Futures Institute, Teesside University, Middlesbrough, UK

Steven W. Singer
Earth Sciences Division, Lawrence Berkeley National Laboratory, Berkeley, CA 94720, USA

Peter Su
Physical Biosciences Division, Lawrence Berkeley National Laboratory, Berkeley, CA 94720, USA; Department of Chemical & Biomolecular Engineering, University of California, Berkeley, CA 94720, USA

Mo Xian
CAS Key Laboratory of Biobased Materials, Qingdao Institute of Bioenergy and Bioprocess Technology, Chinese Academy of Sciences, No.189 Songling Road, Qingdao, Laoshan District 266101, China

Jianming Yang
CAS Key Laboratory of Biobased Materials, Qingdao Institute of Bioenergy and Bioprocess Technology, Chinese Academy of Sciences, No.189 Songling Road, Qingdao, Laoshan District 266101, China

Shang-Tian Yang
Department of Chemical and Biomolecular Engineering, The Ohio State University, USA

Yi-Chun Yeh
Physical Biosciences Division, Lawrence Berkeley National Laboratory, Berkeley, CA 94720, USA; National Taiwan Normal University, Taipei, Taiwan

Haibo Zhang
CAS Key Laboratory of Biobased Materials, Qingdao Institute of Bioenergy and Bioprocess Technology, Chinese Academy of Sciences, No.189 Songling Road, Qingdao, Laoshan District 266101, China

Yanning Zheng
CAS Key Laboratory of Biobased Materials, Qingdao Institute of Bioenergy and Bioprocess Technology, Chinese Academy of Sciences, No.189 Songling Road, Qingdao, Laoshan District 266101, China

Huibin Zou
CAS Key Laboratory of Biobased Materials, Qingdao Institute of Bioenergy and Bioprocess Technology, Chinese Academy of Sciences, No.189 Songling Road, Qingdao, Laoshan District 266101, China

INTRODUCTION

The search for sustainable energy resources is one of this century's great challenges. Biofuels (fuels produced from biomass) have emerged as one of the most promising renewable energy sources, offering the world a solution to its fossil-fuel addiction. They are sustainable, biodegradable, and contain fewer contaminants than fossil fuels.

Although biofuels are rich with promise and may well be a major part of our future energy security, there are still challenges to be met. These will require ongoing investigations in several directions. One direction will be determining more efficient pretreatment technologies. Another fruitful area of research lies in advanced microbial technologies, which can play a major role in more efficient biofuel production. Genetic and chemical engineering research is also required, and nanotechnology has a role to play as well. We need to develop processes and technologies that minimize hydrogen consumption, increase overall process activity, and gain high fuel yields.

This research is crucial to the world's energy needs. The research gathered in this compendium contributes to this vital field of investigation.

—*Juan Carlos Serrano-Ruiz*

Sustainable, economic production of second-generation biofuels is of global importance. This is the basic premise of this book, and in chapter 1, Sekhon and Rahman summarize the facts for us: major technological hurdles remain before we will have widespread conversion of non-food biomass into biofuel. This requires multidisciplinary teams of scientists, technologists, and engineers working together collaboratively to carry out research that will underpin the generation and implementation of sustainable second-generation biofuels.

In Part 2, we move on to specific research. Sabinene, one kind of mono-terpene, accumulated limitedly in natural organisms, is being explored as a potential component for the next generation of aircraft fuels. The demand for advanced fuels impel Rowley and his colleagues to develop biosynthetic routes for the production of sabinene from renewable sugar. In chapter 2, they report their findings that sabinene was significantly produced by assembling a biosynthetic pathway using the methylerythritol 4-phosphate (MEP) or heterologous mevalonate (MVA) pathway combining the GPP and sabinene synthase genes in an engineered *Escherichia coli* strain. Subsequently, the culture medium and process conditions were optimized to enhance sabinene production with a maximum titer of 82.18 mg/L. Finally, the fed-batch fermentation of sabinene was evaluated using the optimized culture medium and process conditions, which reached a maximum concentration of 2.65 g/L with an average productivity of 0.018 g h^{-1} g^{-1} dry cells, and the conversion efficiency of glycerol to sabinene (gram to gram) reached 3.49%. This is the first report of microbial synthesis of sabinene using an engineered *E. coli* strain with the renewable carbon source as feedstock. It establishes a green and sustainable production strategy for sabinene.

Next, in chapter 3, Rana and colleagues investigate bioconversion of lignocellulose by microbial fermentation. This is typically preceded by an acidic thermochemical pretreatment step designed to facilitate enzymatic hydrolysis of cellulose. Substances formed during the pretreatment of the lignocellulosic feedstock inhibit enzymatic hydrolysis as well as microbial fermentation steps. Their review focuses on inhibitors from lignocellulosic feedstocks and how conditioning of slurries and hydrolysates can be used to alleviate inhibition problems. Novel developments in the area include chemical in-situ detoxification by using reducing agents, and methods that improve the performance of both enzymatic and microbial biocatalysts.

Biodiesel and bioethanol, produced by simple and well-known transesterification and fermentationtechnologies, dominate the current biofuel market. However, their implementation in the hydrocarbon-based transport infrastructure faces serious energy-density and compatibility issues. The transformation of biomass into liquid hydrocarbons chemically identical to those currently used in our vehicles can help to overcome these issues eliminating the need to accommodate new fuels and facilitating a smooth transition toward a low carbon transportation system. These strong incentives are favoring the onset of new technologies such as hydrotreating and advanced microbial synthesis, which are designed to produce gasoline, diesel, and jet fuels from classical biomass feedstocks such as vegetable oils and sugars. In chapter 4, Jönsson and his colleagues provide a state-of-the-art overview of these promising routes.

Xylan can be extracted from biomass using either alkali (KOH or NaOH) or dimethyl sulfoxide (DMSO); however, DMSO extraction is the only method that produces a water-soluble xylan. In chapter 5, DMSO extraction of corn stover was studied at different temperatures with the objective of finding a faster, more efficient extraction method. The temperature and time of extraction were compared followed by a basic structural analysis to ensure that no significant structural changes occurred under different temperatures. The resulting data showed that heating to 70 °C during extraction can give a yield comparable to room temperature extraction while reducing the extraction time by ~90 %. This method of heating was shown to be the most efficient method currently available and was shown to retain the important structural characteristics of xylan extracted with DMSO at room temperature.

In chapter 6, my colleagues and I investigated the optimal process conditions leading to high glucose yield (over 80 %) after wet explosion (WEx) pretreatment and enzymatic hydrolysis. The study focused on determining the "sweet spot" where the glucose yield obtained is optimized compared to the cost of the enzymes. WEx pretreatment was conducted at different temperatures, times, and oxygen concentrations to determine the best WEx pretreatment conditions for the most efficient enzymatic hydrolysis. Enzymatic hydrolysis was further optimized at the optimal conditions using central composite design of response surface methodology with respect to two variables: Cellic® CTec2 loading [5 to 40 mg enzyme protein (EP)/g glucan] and substrate concentration (SC) (5 to 20 %) at 50 °C for 72 h. The most efficient and economic conditions for corn stover conversion to glucose were obtained when wet-exploded at 170 °C for 20 min with 5.5 bar oxygen followed by enzymatic hydrolysis at 20 % SC and 15 mg EP/g glucan (5 filter paper units) resulting in a glucose yield of 84 %.

Isoprenoids constitute the largest class of natural products with greater than 55,000 identified members. They play essential roles in maintaining proper cellular function leading to maintenance of human health and plant defense mechanisms against predators, and they are often utilized for their beneficial properties in the pharmaceutical and nutraceutical industries. Most impressively, all known isoprenoids are derived from one of two C_5-precursors, isopentenyl diphosphate (IPP) or dimethylallyl diphosphate (DMAPP). In order to study the enzyme transformations leading to the extensive structural diversity found within this class of compounds there must be access to the substrates. Sometimes, intermediates within a biological pathway can be isolated and used directly to study enzyme/ pathway function. However, the primary route to most of the isoprenoid intermediates is through chemical catalysis. In chapter 7, Jarchow-Choy and her colleagues provide a thorough examination of synthetic routes to isoprenoid and isoprenoid precursors with particular emphasis on the syntheses of intermediates found as part of the 2C-methylerythritol 4-phosphate (MEP) pathway. In addition, representative syntheses are presented for the monoterpenes (C_{10}), sesquiterpenes (C_{15}), diterpenes (C_{20}), triterpenes (C_{30}) and tetraterpenes (C_{40}). Finally, in some instances, the synthetic routes to substrate analogs found both within the MEP pathway and downstream isoprenoids are examined.

In chapter 8, Yang and Liu introduce the focus of Part 4: the role of genetic engineering in the new biotechnologies. Metabolic process engineering (MPE) is a powerful technology that integrates the well-developed process control techniques, such as precise bioreactor controllers and in situ sensors, and advanced omics technologies. It enables the rational design of a bio-production process, and thus can lead to a highly efficient fermentation process for biochemicals and biofuels production. Different from the well-known traditional fermentation process development, MPE targets to engineer the bio-production process by controlling the cell physiology and metabolic responses to changes in fermentation process parameters and incorporating the interplay between cell and process into the rational process design. Yang and Liu focus on the application of MPE to improve biochemicals and biofuels production via precise bioreactor controllers, in situ sensors, and omics technologies.

Transgenic microalgae have the potential to impact many diverse biotechnological industries, including energy, human and animal nutrition, pharmaceuticals, health and beauty, and specialty chemicals. However, the lack of well-characterized transformation vectors to direct engineered gene products to specific subcellular locations, and the inability to robustly express multiple nuclear-encoded transgenes within a single cell have been major obstacles to sophisticated genetic and metabolic engineering in algae. In chapter 9, Rasala and her colleagues validate a set of genetic tools that enable protein targeting to distinct subcellular locations, and present two complementary methods for multigene engineering in the eukaryotic green microalga *Chlamydomonas reinhardtii*. The tools described will enable advanced metabolic and genetic engineering to promote microalgae biotechnology and product commercialization.

The chemoautotrophic bacterium *Ralstonia eutropha* can utilize H_2/CO_2 for growth under aerobic conditions. While this microbial host has great potential to be engineered to produce desired compounds (beyond polyhydroxybutyrate) directly from CO^2, little work has been done to develop genetic part libraries to enable such endeavors. In chapter 10, Bi and colleagues report the development of a toolbox for the metabolic engineering of *Ralstonia eutropha* H16. They have constructed a set of broad-host-range plasmids bearing a variety of origins of replication, promoters, 5' mRNA stem-loop structures, and ribosomal binding sites. Specifically, they analyzed the origins of replication pCM62 (IncP), pBBR1, pKT (IncQ), and their variants. They tested the promoters PBAD, T7, Pxyls/PM, PlacUV5, and variants thereof for inducible expression. They also evaluated a T7 mRNA stem-loop structure sequence and compared a set of ribosomal binding site (RBS) sequences derived from *Escherichia coli, R. eutropha*, and a computational RBS design tool. Finally, the authors employed the toolbox to optimize hydrocarbon production in *R. eutropha* and demonstrated a 6-fold titer improvement using the appropriate combination of parts. They constructed and evaluated a versatile synthetic biology toolbox for *Ralstonia eutropha* metabolic engineering that could apply to other microbial hosts as well.

Heterogeneous photocatalysis has become a comprehensively studied area of research during the past three decades due to its practical interest in applications including water–air depollution, cancer therapy, sterilization, artificial photosynthesis (CO_2 photoreduction), anti-fogging surfaces, heat transfer and heat dissipation, anticorrosion, lithography, photochromism, solar chemicals production, and many others. The utilization of solar irradiation to supply energy or to initiate chemical reactions is already an established idea. Excited electron–hole pairs are generated upon light irradiation of a wide-band gap semiconductor, which can then be applied to solar cells to generate electricity or in chemical processes to create/degrade specific compounds. While the field of heterogeneous photocatalysis for pollutant abatement and mineralisation of contaminants has been extensively investigated, a new research avenue related to the selective valorization of residues has recently emerged as a promising alternative to utilize solar light for the production of valuable chemicals and fuels. The review in chapter 11 focuses on the potential and applications of solid photonanocatalysts for the selective transformation of biomass-derived substrates.

In chapter 12, Das and El-Safty explore the nanoassembly of nearly monodisperse nanoparticles (NPs) as uniform building blocks to engineer zirconia (ZrO_2) nanostructures with mesoscopic ordering by using a template as a fastening agent. They investigate mesophase of the materials through powder X-ray diffraction and TEM analysis (TEM) and N_2 sorption studies. The TEM results revealed that the mesopores were created by the arrangement of ZrO_2 NPs with sizes of 7.0–9.0 nm and with broad interparticle pores. Moreover, the N_2 sorption study confirmed the results. The surface chemical analysis was performed to estimate the distribution of Zr, O, and S in the sulfated ZrO_2 matrices. The materials in this study displayed excellent catalytic activity in the biodiesel reaction for effective conversion of long-chain fatty acids to their methyl esters, and the maximum biodiesel yield was approximately 100%. The excellent heterogeneous catalytic activity could be attributed to the open framework, large surface area, presence of ample acidic sites located at the surface of the matrix, and high structural stability of the materials. The catalysts revealed a negligible loss of activity in the catalytic recycles.

Finally, in chapter 13, Feyzi and colleagues investigate the transesterification reaction over the $PW_{12}O_{40}/Fe-SiO_2$ catalyst prepared, using sol-gel and impregnation procedures in different operational conditions. Experimental conditions were varied as follows: reaction temperature 323–333 K, methanol/oil molar ratio = 12/1, and the reaction time 0–240 min. The $H_3PW_{12}O_{40}$ heteropolyacid has recently attracted significant attention due to its potential for application in the production of biodiesel, in either homogeneous or heterogeneous catalytic conditions. Although fatty acids esterification reaction has been known for some time, data is still scarce regarding kinetic and thermodynamic parameters, especially when catalyzed by nonconventional compounds such as $H_3PW_{12}O_{40}$. This kinetic study utilizing Gc-Mas in situ allows for evaluating the effects of operation conditions on reaction rate and determining the activation energy along with thermodynamic constants including G, S, and H. The authors' results indicate that the $PW_{12}O_{40}$/Fe-SiO$_2$ magnetic nanocatalyst can be easily recycled with a little loss by magnetic field and can maintain higher catalytic activity and higher recovery even after being used five times.

PART I

THE PREMISE

CHAPTER 1

Synthetic Biology: A Promising Technology for Biofuel Production

KAMALJEET KAUR SEKHON AND PATTANATHU K. S. M. RAHMAN

With the increasing awareness among the masses and the depleting natural resources and oil reservoirs, a replacement for the fossil fuels is urgently required. There are rising global concerns about climate change and energy security. The current biofuel production trends are no doubt promising and increasing steadily. The biofuel markets are getting bigger and better in the European Union, USA, Brazil, India, China and Argentina and contributing to their bio-economies, respectively. In the US, biodiesel production exceeded 1 billion gallons in 2012 and reports claim that the global biofuels market will touch the figure of $185 billion in 2021. However the big question is: will the current biofuel production rate be able to meet the escalating transportation fuel demands?

Synthetic Biology: A Promising Technology for Biofuel Production. © 2013 Sekhon KK, et al. Journal of Petroleum and Environmental Biotechnology 4:*e121. doi: 10.4172/2157-7463.1000e121. Creative Commons Attribution License. Used with permission of the authors.*

The total estimated generation of biomass in the world is 150 billion tons annually. Increase in the production of biofuels in the recent years and the usage of edible commodities like maize, sugarcane and vegetable oil has led to the worldwide apprehension towards the future of biofuels and to the 'food vs fuel' debate. The second generation biofuels, however, are produced from renewable, cheap and sustainable feed-stocks for example citrus peel, corn stover, sawdust, bagasse, straw, rice peel and are attracting ever-increasing attention. A great deal of research, by the scientific community, is carried out in various parts of the world in order to improve the yield of second generation biofuels to meet the future demands but hasn't achieved any remarkable success.

Sustainable, economic production of second generation biofuels is of global importance. However, major technological hurdles remain before widespread conversion of non-food biomass into biofuel. Various multi-disciplinary teams of scientists, technologists and engineers work together collaboratively in integrated teams to carry out research that underpins the generation and implementation of sustainable second generation biofuels from algal biomass using biological processes. The advantages of algal biomass from both micro and macroalgae as a raw material for producing biofuels have been well recognised for decades. Billions of tonnes of algal biomass are enzymatically converted into food energy by marine and freshwater animals and microbes every day, in a sustainable manner. However, the industrial, enzyme driven conversion of such biomass for bioenergy applications is still in its infancy.

The production of commercially attractive biofuels using enzymatic methods, all the same, is not as easy as it appears. The various polysaccharides viz. cellulose, starch, lignin, hemicellulose, or lignocelluloses need to be enzymatically degraded for their transformation into glucose or sugar molecules which in turn are fermented into biofuels (bioethanol or biobutanol). In case of cellulose, the process of cellulolysis involves enzymes like cellulases and glucosidases. Cellulases are expensive, unstable and slow in action; therefore they increase the overall economics of the process of cellulolysis and hence biofuel production. The bulk production of cellulases at industrial level seems to be the relevant solution. The microbes that produce cellulases include symbiotic anaerobic bacteria (e.g. *Cellulomonasfimi, Clostridium thermocellum, Clostridium phytofermentans, Thermobifidafusca*) found in ruminants such as cow and sheep, flagellate protozoa present in hinduts of termites, and filamentous fungi isolated from decaying plants (e.g. *Hypocreajecorina, Thermoascusaurantiacus, Phanerochaetechrysosporium, Neurosporacrassa, Tricodermareesei, Asperigillusniger, Fusariumoxysporum*). The gene(s) responsible for cellulase production are characterized, isolated and recombinantly introduced into *Escherichia coli* for the enhanced cellulase expression levels.

Apart from the conventional biotechnology methods for biofuel production, synthetic biology has shown promising results lately. Understanding the DNA sequences, precisely measuring the gene behaviour paves way for fabricating or synthesizing the cellulase gene *de novo*. To put it in simple words, synthetic biology is a science of designing and constructing new biological parts, devices and systems for programming cells and organisms and endowing them with novel functions. It is a technique of writing the DNA / genetic code base by base using several computational tools and software's like Gene designer, GenoCAD, Eugene and Athena to name a few. Gene designer is a DNA design tool for *de novo* assembly of genetic constructs, GenoCAD is a computer-assisted-design application for synthetic biology for designing complex gene constructs and artificial gene networks, Eugene is a language designed to develop novel biological devices and Athena is a CAD / CAM software for constructing biological models as modules. These synthetic biology approaches can be useful in bringing down the cost of cellulases and, thereby, of biofuels. Several companies are spending a fortune on the production of bioethanol for example; Amyris Biotechnologies, Verenium, Iogen, Bioethanol Japan, Mascoma, POET, SolixBiofuels, Pacific Ethanol, NextGen Fuel Inc. and Jatro Diesel. However, the cost-effective production of the second generation biofuels is still a cherished desire of the scientific community.

Synthetic biology is an evolving field still dealing with the inherent complexity of biological systems and overcoming the biosafety issues involved with engineering the living systems. Indeed the proliferation of the computer modelling tools is leading to the revolution of this discipline which might write the success story of some of the present and future scientific challenges.

PART II

PRETREATMENT TECHNOLOGIES

CHAPTER 2

Efficient Extraction of Xylan from Delignified Corn Stover Using Dimethyl Sulfoxide

JOHN ROWLEY, STEPHEN R. DECKER, WILLIAM MICHENER, AND STUART BLACK

2.1 INTRODUCTION

Biofuels are becoming more widespread throughout the United States as more advanced conversion methods become available. The most advanced process currently is the conversion of lignocellulosic biomass into ethanol (Kim et al. 2009). Despite having much larger production potential than starch-based ethanol, lignocellulosic ethanol is still in the early stages. The conversion of biomass sugars into biofuels is an important aspect of the Department of Energy's mission to promote the integration of renewable fuels and is a key component in the worldwide move towards renewable energy. Before additional progress can be made, it is desirable to understand in detail the mechanisms that occur during the biomass to biofuel conversion process.

Efficient Extraction of Xylan from Delignified Corn Stover Using Dimethyl Sulfoxide. Copyright © The Author(s) 2013. 3 Biotech. 2013 Oct; 3(5): 433–438. doi: 10.1007/s13205-013-0159-8. Creative Commons Attribution License (http://creativecommons.org

Biomass is made up of three components: cellulose, hemicellulose and lignin. Xylan, a prevalent plant cell wall polymer made up of mostly xylose, is of particular interest as the dominant plant cell wall hemicellulose (Ebringerová et al. 2005). One of the challenges associated with the efficient production of biofuels involves the selective removal and/or hydrolysis the polymeric xylose backbone of xylan. During neutral or acidic thermochemical pretreatment of biomass, xylan is removed from the biomass and broken down into xylose, arabinose, and a few other minor components such as acetic acid (Naran et al. 2009).

To better understand the mechanism of thermochemical and enzymatic removal of xylan, it is useful to develop antibodies capable of tagging xylan in biomass. Antibodies can be tagged with fluorescent dyes, allowing the location of the xylan in biomass to be tracked either optically or spectrophotometrically prior to and following pretreatment. By identifying the location of the xylan, the pretreatment process and the subsequent fermentation process can be tailored to improve ethanol production. Antibody tagging can be very beneficial in understanding the mechanism of xylan removal, however, to create specific antibody tags, a native-like xylan is desirable. Many extraction methods result in degradation or deacetylation of the xylan resulting in a non-native, water-insoluble product, which could potentially produce antibodies with non-useful specificity, as specific side groups are missing. Dimethyl sulfoxide (DMSO) extractions have been found to result in a water-soluble form of xylan, which retains the acetyl groups present in the native state (Hägglund et al. 1956). This native-like xylan is more likely to result in production of antibodies specific to the native structures found in xylans in situ in the cell wall.

In this study, a DMSO extraction of xylan in corn stover was studied at varying temperatures of extraction to determine an ideal temperature for efficient extraction.

When extracting xylan from biomass with DMSO, a pretreatment of the sample is necessary to open the cell structure and allow the polymeric xylans freedom to be extracted. Owing to the coupling between xylan and lignin, xylan is intractable until much of the lignin has been removed or these connections severed. Decoupling of xylan from lignin is important in accessing xylan in biomass, but complete removal of lignin will result in loss of xylan from the sample (Ebringerová et al. 2005). Multiple delignification procedures exist for the removal of lignin from corn sto-ver, however, acid-chlorite bleaching was found to be the most efficient method of delignification without excessive de-acetylation of the xylan (Ebringerová et al. 2005).

Following delignification, xylan is extracted from the sample. Often xylan is extracted with KOH or NaOH (Ebringerova and Heinze 2000). However, this method results in de-esterification of the acetyl groups present on the xylan (via saponification of the ester links), leading to a water insoluble product which has limited utility for antibody production and as a substrate for hemicellulase assays. Therefore, in this study, xylan was removed by DMSO extraction to retain the acetyl groups, resulting in a water-soluble product. The extraction was first performed at room temperature, following the method proposed by Hägglund et al. (1956) in 1956. This method is carried out by stirring the biomass in DMSO for approximately 24 h at room temperature. A series of extractions was then performed at higher temperatures (70 °C and at 40 °C) with variable times of extraction. The yields resulting from the extractions were compared and, including the time required to perform each extraction, the most ef-ficient method of extraction was determined.

Further analysis was performed on each sample to determine the con-tent of the yield acquired through extraction and to ensure that no sig-nificant structural changes took place under heated conditions. Infrared spectroscopy and QToF MS analysis was used to determine the general structural features and to ensure that no de-esterification or de-polymer-ization took place during the heated extractions.

2.2 METHODS

2.2.1 DELIGNIFICATION OF BIOMASS

Approximately 300 g of milled corn stover was extracted in a polypro-pylene thimble using a Soxhlet extractor following NREL's Determina-tion of Biomass Extractives Laboratory Analytical Procedure (Sluiter et al. 2008). The NREL procedure is a two-step procedure carried out in a Soxhlet extractor. All extractions are carried out at the reflux temperature of the solvent used and at ambient pressure. Each extraction is performed until little to no color is present in the extraction chamber. Depending on the nature of the material, this takes between 18 and 48 h for each step. The first extraction was performed with de-ionized (DI) water to remove accessible water-soluble compounds. A second extraction was performed using ethanol to remove lipids and other extractables. The solid sample was air-dried following ethanol extraction prior to delignification.

Delignification was carried out in double bagged one gallon plastic zipper closure bags by adding water to the approximately 100 g of air-dried, extracted biomass at a biomass/water consistency of 10 %. Approximately, 40 g of sodium chlorite ($NaClO_2$) was added to the mixture and the bag was mixed well followed by a 5 mL addition of concentrated hydrochloric or glacial acetic acid. A smaller volume of hydrochloric acid is needed to sustain the reaction. The bag was closed and heated in a 60 °C water bath in a fume hood for approximately 3 h. Regular venting of the bag was required to relieve pressure in the bag and prevent reaching too high a concentration of ClO_2. If the concentration of chlorine dioxide in the atmosphere of the bag or any bleaching vessel is too high, a "puff" can result from the decomposition of the chlorine dioxide. A "puff" is a term coined within the pulping industry to differentiate a low speed detonation wave of <1 m/s from an explosion wave (>300 m/s). Plastic zipper bags will open in the event of a puff releasing the gas without creating a debris hazard (Fredette 1996).

Once every hour, an additional 40 g of $NaClO_2$ was added to the bag until the total amount was approximately 0.70 g $NaClO_2$/g biomass. The remaining liquid was filtered from the solids and the solid biomass was thoroughly washed with DI water and lyophilized prior to DMSO extraction.

2.2.2 DMSO EXTRACTION

A 1 L electrically heated reaction flask fitted with an overhead mechanical stirrer was used for all extractions. Approximately 50 g of delignified corn stover was added to a flask and extracted with DMSO using a ratio of approximately 14 mL/g biomass at room temperature with stirring at 20 rpm for a specified time. The solid was filtered and extracted a second time with DMSO for the same time period. The solid was filtered and washed thoroughly with ethanol to remove residual DMSO and extracted xylan. The ethanol filtrate was reserved for the precipitation step. The DMSO extracts were combined and absolute ethanol was added to the DMSO extract (3.8 L ethanol/L of final extract). Concentrated hydrochloric acid (HCl) was added in a ratio of approximately 0.66 mL HCl/L of ethanol/ DMSO solution to precipitate the xylan from the DMSO/ethanol mixture. The solution was cooled at 4 °C overnight to complete precipitation. The cold solution was filtered though paper filter (Whatman Grade 1). The filter paper and isolated xylan were macerated, washed with ethanol and stirred overnight in a small amount of ethanol. Ethanol was filtered from the solid xylan and macerated paper filter. The resulting filter cake was stirred overnight with fresh ethanol to remove as much DMSO as possible and filtered. The filter cake was further washed with diethyl ether with overnight stirring to remove any remaining ethanol and DMSO. The xylan was dissolved away from the macerated paper fibers in warm water (30 °C), filtered with small amounts of water added for washing and lyophi- lized.

The DMSO extraction was carried out at 20, 40 and 70 °C according to the conditions shown in Table 1. Extractions at 40 and 70 °C were per- formed in duplicate. Extraction at 20 °C was a single extraction.

TABLE 1: The conditions for four subsequent extractions at temperatures above room temperature.

Temperature (°C)	Time (h)	Number of extractions
70	2	2
70	2	1
70	1	2
40	4	2
30	24	2

The time noted above represents the duration of each extraction

2.2.3 SAMPLE ANALYSIS

The final products were analyzed qualitatively by their water solubility and for yield from the bleached material by mass. The methods were compared according to yield and time efficiency. Samples were analyzed on a Thermo Scientific Nicolet 6700 FTIR Spectrometer fitted with a Smart iTR diamond cell and a DTGS detector. Samples were scanned for 150 scans and compared to previously isolated and analyzed samples (Ebringerová et al. 2005).

Two samples, one extracted at room temperature, and the other at 70 °C, were prepared in a 50/50 solution of H2O/acetonitrile in 0.2 % formic acid. Each sample was directly infused into a Micromass Q-ToF micro (Micromass, Manchester, UK) quadrupole time of flight mass spectrometer with a 250 μL Hamilton gastight syringe (Hamilton, Reno, NV, USA) at a flow rate of 5 μL/min. Spectra were obtained in positive MS mode from a mass range of 600–1,500 m/z and processed by Masslynx data system software (Micromass, Manchester, OK). In positive-ion MS mode, cone voltage was set at 30 volts and capillary at 3,000 volts. Both cone and desolvation gas flows were optimized at 10 and 550 L/h, respectively. Source and desolvation temperatures were set at 100 and 250 °C. Sample mass spectra were collected for 2 min to ensure adequate signal levels. Mass calibration was performed using a solution of 2 pmol/μL of sodium rubidium iodide solution. The calibration mix was collected for 2 min and summed.

2.3 RESULTS

Figure 1 shows the percent yield for each of the extracted samples. Analysis of the yield for the different methods indicates that heating during the extraction process results in no significant loss of recovery. It is also evident that much less time is needed for extraction when the DMSO is heated during extraction compared to a room temperature extraction. At room temperature, two sequential extractions, each lasting a full 24 h, are necessary and resulted in a xylan/delignified biomass yield of 8.7 %. Upon heating to 70 °C and decreasing the extraction time to only 2 h, the yield was 8.6 ± 0.2 %. Even when the extraction time was decreased to 1 h for a sample heated to 70 °C, the loss in yield was not found to be particularly large (7.6 ± 0.6 %). However, a significant loss in yield was found when the sample was extracted only once with DMSO. The percent yield did not drop significantly when the sample was heated to 40 °C, but a longer extraction time was necessary (Fig. 1).

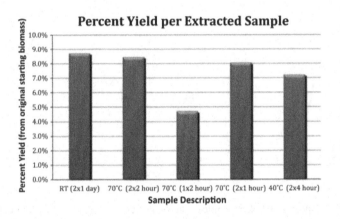

FIGURE 1: % yield of xylan extracted from bleached material. Descriptors of each sample are temperature (# replicates x time of extraction for each replicate); RT room temperature.

By infrared spectroscopy, there is little structural difference between the heated and room temperature extractions. Figure 2 compares corn stover xylan extracted using DMSO and a commercial oat spelt xylan (Fluka) extracted under alkaline conditions. The commercial xylan has no signal for the acetate ester present in the DMSO xylans at ~1,700 and ~1,300/cm which are the well-known carbonyl and ester linkage absorbance bands for the acetyl groups. The commercial xylan shows a slight absorbance at 1,500/cm which is indicative of residual lignin. The DMSO extracted lignin does not have an absorbance in this region. This would indicate that no either no residual lignin was present following acid chlorite delignification or that no water soluble lignin was present in the isolated xylan following lyophilization.

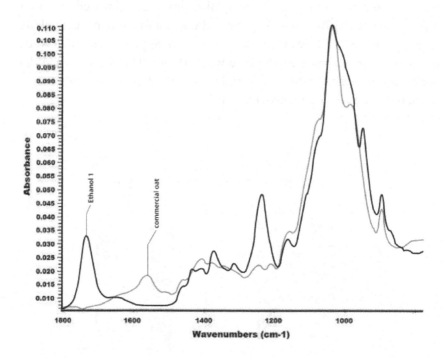

FIGURE 2: IR-spectra of xylan samples. The *black spectrum* is of the DMSO extracted xylan. The *gray spectrum* is of a commercial oat xylan sample extracted with alkaline conditions. The peaks at 1,735 and 1,235/cm are indicative of carbonyl and ester linkage, respectively, of the acetyl groups on the xylan polymer.

When comparing (Fig. 3) the two DMSO extracted corn stover xylans, it is clear that heating during the DMSO extracting process does not influence the structure in a significant way. The expected peaks are present for the DMSO extracted sample indicating the presence of an ester group.

Both the room temperature- and 70 °C-extracted samples were water soluble, providing further evidence of the presence of acetyl groups on the isolated xylans, as acetylation is known to provide for water solubilization of xylans (Grondahl and Gatenholm 2005; Gabrielii et al. 2000). Figure 4 shows the mass spectra comparison between xylan extracted at room temperature and xylan extracted at 70 °C. The MS spectra collected from each sample shows a degree of polymerization range of 4–9 residues, indicative

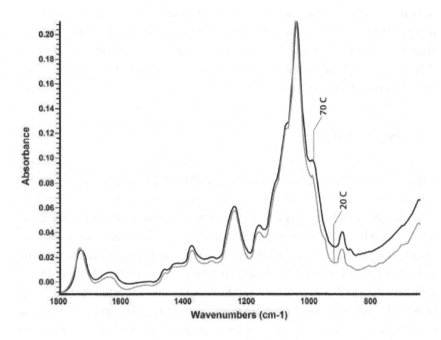

FIGURE 3: IR-spectra of two extracted xylan samples. The *black line* is the spectrum of the DMSO extracted sample extracted at 70 °C. The *gray line* is the spectrum for a corn stover extracted xylan at room temperature.

of the limitations of the ESI technique, rather than the actual DP of the sample materials. Low MW xylo-oligomer standards (DP 2–4) showed no fragmentation at the voltages used in this study (data not shown), indicating that the ions detected in the samples are generated during the sample preparation and remain with the soluble fraction during purification, not as artifacts of MS fragmentation. There are slight differences in the two spectra, specifically with intensities seen at varying masses with the relative abundance. The spectra show that a high number of the masses associated with each sample are present in the other. It is clear from the fragments in the spectra that the two xylan samples are structurally very similar, supporting the analysis from the IR instrument, and showing that no significant structural changes occurred when xylan was extracted at a higher temperature. Elucidation of the structure of the isolated xylans may be found in our previous work (Naran et al. 2009) and was not attempted in this study, which is primarily aimed at developing a faster, easier method for obtaining native xylans.

2.4 DISCUSSION

Drawing from these results, it can be concluded that heating during an extraction can increase the efficiency of a xylan extraction. A heated extraction requires much less time than an extraction done at room temperature. It also must be mentioned that while further study is needed, these preliminary results predict that the number of extractions (proportional to the total volume of DMSO used) does impact that percent yield. The conclusion that percent yield is increased upon heating is not supported by this study, however, further analysis can be done to confirm this prediction. These results strongly indicate that the yield obtained under heated conditions is comparable to that of an unheated extraction and requires significantly less time to extract (~9 % of the total time required to extract an unheated sample). Further, it must also be said that heating during the extraction process within the temperature range studied here does not change the xylan structure. The product is not de-esterified during the process and remains water soluble. Furthermore, no significant or obvious structural changes were observed when comparing heated samples to non-heated samples (Figs. 3, 4).

The efficiency of xylan extraction with DMSO can be greatly improved if the samples are heated during extraction. The percent yield resulting from an extraction performed at 70 °C for 2 h with multiple extractions is comparable to the yield at room temperature with two 24 h extractions. Provided that no de-esterification of the xylan results from heating the sample as shown in Fig. 3, heating increases the efficiency of the extraction. From this study, it was determined that the most efficient method of extraction is the following: two 70 °C extractions each lasting 2 h. This method provided a yield of 8.6 ± 0.2 % which is considered to be sufficient for the purposes of this study

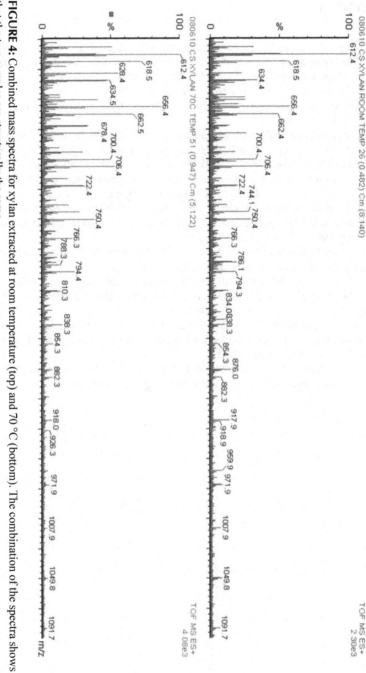

FIGURE 4: Combined mass spectra for xylan extracted at room temperature (top) and 70 °C (bottom). The combination of the spectra shows that the two samples are structurally the same.

REFERENCES

1. Ebringerova A, Heinze T (2000) Xylan and xylan derivatives—biopolymers with valuable properties, 1—naturally occurring xylans structures, procedures and properties. Macromol Rapid Comm 21(9):542–556. doi:10.1002/1521-3927(20000601)21: 9<542::Aid-Marc542>3.3.Co;2-Z

2. Ebringerová A, Hromadkova Z, Heinze T (2005) Hemicellulose. Adv Polym Sci 186:1–67

3. Fredette MC (1996) Bleaching chemicals: chlorine dioxide. In: Reeve DW, Dence CW (eds) Pulp bleaching: principles and practice. Tappi Press, Atlanta, pp 67–68

4. Gabrielii I, Gatenholm P, Glasser WG, Jain RK, Kenne L (2000) Separation, characterization and hydrogel-formation of hemicellulose from aspen wood. Carbohydr Polym 43(4):367–374. doi:10.1016/S0144-8617(00)00181-8

5. Grondahl M, Gatenholm P (2005) Role of acetyl substitution in hardwood xylan. In: Dumitriu S (ed) Polysaccharides: structural diversity and functional versatility. Marcel Dekker, New York, pp 509–514

6. Hägglund E, Lindberg B, McPherson J (1956) Dimethylsulphoxide, a solvent for hemicelluloses. Acta Chem Scand 10:1160–1164

7. Kim TH, Nghiem NP, Hicks KB (2009) Pretreatment and fractionation of corn stover by soaking in ethanol and aqueous ammonia. Appl Biochem Biotechnol 153(1–3):171–179. doi:10.1007/s12010-009-8524-0

8. Naran R, Black S, Decker SR, Azadi P (2009) Extraction and characterization of native heteroxylans from delignified corn stover and aspen. Cellulose 16(4):661–675. doi:10.1007/S10570-009-9324-Y

9. Sluiter A, Ruiz R, Scarlata C, Sluiter J, Templeton D (2008) Determination of extractives in biomass. National Renewable Energy Laboratory, Golden.

CHAPTER 3

Process Modeling of Enzymatic Hydrolysis of Wet-Exploded Corn Stover

VANDANA RANA, DIWAKAR RANA, AND BIRGITTE K. AHRING

3.1 INTRODUCTION

Corn stover is abundantly available in the Midwest of USA and can be an excellent feedstock for biofuel production because of its lower lignin content compared to woody biomass. The commercial viability of biorefineries based on corn stover has been burdened by the use of expensive enzymes needed to hydrolyze the biomass material after pretreatment [1, 2]. It has been well established that producing higher concentration of sugars is an absolute necessity in an industrial setting as it lowers the heating requirements (lowering operating cost) and increases the volumetric efficiency (lowering capital cost) of the equipment [3]. Therefore, lowering the enzyme input and increasing the dry matter content during enzyme hydrolysis for higher cellulose conversion would be one of the most significant steps towards the direction of bioethanol production cost reduc-

Process Modeling of Enzymatic Hydrolysis of Wet-Exploded Corn Stover. © *2013 Jonsson et al.; licensee BioMed Central Ltd. Biotechnology for Biofuels 2013, 6:16 (doi:10.1186/1754-6834-6-16). Creative Commons Attribution License (http://creativecommons.org/licenses/by/2.0).*

tion and eventually leading to the commercialization of second-generation biorefineries based on the lignocellulosic feedstock.

Several researchers have worked on using corn stover for bioethanol production. Kaar et al. [4] used lime pretreatment followed by enzymatic hydrolysis at 5 % substrate concentration (SC) and 20 filter paper units (FPU) (Spezyme CP and Novozym 188) and obtained 60 % cellulose conversion. Kim et al. [5, 6] introduced ammonia recycle percolation pretreatment followed by enzyme hydrolysis at 1 % SC and 10 FPU [Spezyme CP and β-glucosidase (Sigma)] and obtained 92 % cellulose conversion. This concentration of solids will, however, be far from an industrial process. Bura et al. [7] used SO_2-catalyzed pretreatment followed by enzyme hydrolysis at 8 % SC and 10 FPU (Spezyme CP, Novozym 188, and Multifect® Xylanase) and obtained 100 % cellulose conversion. However, again, the solid concentration was far lower than that needed for operating any industrial process. Using chemicals such as sulfur could further affect the downstream processing of products; for instance, sulfur will be attached to the solid fraction that remained after sugar extraction [8]. Recently, Yang et al. [9] used steam explosion pretreatment followed by enzyme hydrolysis at 25 % SC and 20 FPU (Celluclast) and obtained 85 % cellulose conversion. Even though this study achieved high glucose concentrations, the amount of enzymes used was higher, affecting the applicability of the process.

In this work, we will investigate the most efficient and cost-effective process conditions for producing sugars from corn stover using wet explosion (WEx) pretreatment, which only require oxygen and heat to open the structure of the biomass and which can operate at high SC [10, 11]. Our focus will be on finding the optimal conditions producing high yields of glucose while keeping the enzyme dose low—the "sweet spot." As a first step, screening of WEx pretreatment will be performed to find the best pretreatment condition, which will be used for the optimization of enzymatic hydrolysis. The second step will be central composite design (CCD) to generate response surface, model fitting, and prediction of optimum parameters for enzymatic hydrolysis of WEx corn stover. The third step will be a scale-up study of enzymatic hydrolysis at optimized condition from lab scale to 100-L pilot scale. In addition, we will evaluate the process economics for determining the minimum enzyme dose, which still can support a biorefinery with corn stover as raw material.

3.2 MATERIALS AND METHODS

3.2.1 PREPARATION OF BIOMASS

Corn stover with a particle size of approximately quarter inch was kindly provided by Iowa State University, Ames, IA, USA. Before pretreatment, corn stover was milled to 2-mm particle size with a Retsch cutting mill SM 200 (Retsch Inc., PA, USA) and kept at room temperature. A portion of corn stover was milled to 1-mm particle size for the compositional analysis. The compositional analysis was performed in duplicates. The moisture content of the raw corn stover was found to be 8 %.

3.2.2 EXPERIMENTAL APPROACH

Figure 1 shows the experimental setup used in the present study. The milled corn stover was pretreated using WEx at 12 different pretreatment conditions with varying temperatures (160 to 180 °C), residence time (15 to 25 min), and oxygen pressure (2 to 6 bar). In order to investigate the efficiency of pretreatment, enzymatic digestibility test was conducted with Cellic® CTec2 (40 mg enzyme protein (EP)/g glucan) on all 12 pretreatment slurries. Sugars, sugar degradation products, and other fermentation inhibitors [hydroxymethylfurfurals (HMFs), furfural, and acetate] released during pretreatment or after enzymatic hydrolysis were analyzed using high-performance liquid chromatography (HPLC). Based on the sugar yield and the concentration of inhibitors, the optimal pretreatment condition was determined and used for finding the minimal enzyme dose needed for obtaining at least 80 % of the potential cellulose conversion from the biomass. This selected pretreatment was referred as the platform WEx pretreatment.

The enzymatic hydrolysis study to optimize the enzyme dosage was performed in shake flasks on the platform WEx-pretreated sample using

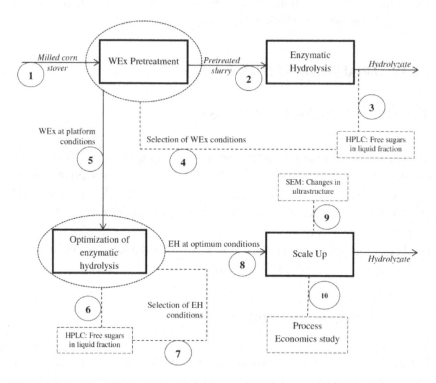

FIGURE 1: Schematic of experimental methodology. The line number is shown within the circle and indicates the step order.

varying SCs (5 to 20 %) and enzyme loading (5 to 40 mg EP/g glucan). Test of two parameters resulted in 14 experimental runs. The amount of free sugars released after these 14 enzymatic hydrolysis runs was analyzed using HPLC. Response surface methodology (RSM) was adopted to determine the optimal hydrolysis condition. Finally, the results for the optimal enzyme dose study were checked in a 100-L pilot-scale test at the optimal conditions (SC and enzyme loading) and the effect on the ultrastructure was followed using scanning electron microscopy to evaluate the changes done to the biomass during enzymatic hydrolysis. Finally, we analyzed the process economics to substantiate the effect of optimization study on the overall cost reduction of ethanol.

3.2.3 MEASUREMENT OF PROTEIN ASSAY AND CELLULASE ACTIVITY

The protein concentration of Cellic® CTec2 (kindly provided by Novozymes, Franklinton, NC, USA) was determined using standard Pierce® BCA Protein Assay Kit (Thermo Scientific, Waltham, MA, USA). The protein concentration in Cellic® CTec2 was measured as 210 mg/mL of enzyme. Filter paper activity of Cellic® CTec2 was measured according to enzyme activity assay [12]. The cellulolytic activity of concentrate was determined as 74 FPU/mL.

3.2.4 SELECTION OF PLATFORM WEx PRETREATMENT PARAMETERS

3.2.4.1 WEx PRETREATMENT OF CORN STOVER

Corn stover was pretreated by WEx in a pilot plant at Washington State University as previously described [13]. The 12 different conditions used for the WEx pretreatment are shown in Table 1. Corn stover samples resulting from these pretreatment conditions are designated as CS-1 through CS-12. The 10-L WEx reactor was loaded with raw corn stover and water in proportions resulting in a total slurry of 5 kg with 25 % SC. After closing the reactor, oxygen was introduced into the reactor at pressures ranging from 2 to 6 bar. After introducing the oxygen, the pretreatment reactor was heated to temperatures varying from 160 to 180 °C with residence times at the desired temperature between 15 and 25 min (heating time, about 5 min). At the end of pretreatment reaction time, the pretreated biomass was flashed into a 100-L flash tank. The slurry was used for the optimization of enzymatic hydrolysis experiment ("Hydrolysis Testing"), and a small portion of the slurry was kept for solid and liquid analysis.

TABLE 1: Pretreatment conditions for WEx of 25 % DM used in the present study.

Sample number	Reaction temperature (°C)	Residence time (min)	Oxygen pressure (bar)
CS-1	160	15	4
CS-2	160	20	2
CS-3	160	20	6
CS-4	160	25	4
CS-5	170	15	2
CS-6	170	15	6
CS-7	170	20	4
CS-8	170	20	5.5
CS-9	170	25	2
CS-10	180	15	4
CS-11	180	20	2
CS-12	180	20	6

3.2.4.2 HYDROLYSIS TESTING

The pretreated slurries from 12 different runs were hydrolyzed using Cellic® CTec2 with 40 mg EP/g glucan to investigate the maximum digestibility. A total of 50 g of well-mixed pretreated slurry with 25 % SC was taken for enzymatic hydrolysis. The enzymatic hydrolysis of the pretreated samples was performed at pH 5 and 50 °C for 72 h in 125-mL shake flasks. In order to overcome the mixing challenges of 25 % SC in the Erlenmeyer shake flasks, the pretreated samples were diluted to 20 % SC using 0.1 M sodium citrate buffer. Maximum care was taken to ensure that the solid/liquid ratio was maintained during sampling using intense mixing. The pretreated sample with the highest sugar concentration and the minimal concentration of sugar degradation products was selected for the enzymatic hydrolysis optimization studies. Experiment for the enzyme optimization was then designed as described in "Optimization of Enzymatic Hydrolysis."

The glucose yield is determined by Eq. (1) as described below:

$$\%Yield = \frac{(Glu_{EH,L})}{1.111 \times F_{celluloseRB} \times (ini.sol)} \times 100 \quad (1)$$

where (GluEH,L) is the glucose concentration (grams per liter) in enzymatic hydrolyzate liquid, FcelluloseRB is the fraction of cellulose in the raw biomass as determined by compositional analysis, and (ini.sol) is the initial solid concentration (grams per liter) at which the enzymatic hydrolysis was performed.

3.2.5 OPTIMIZATION OF ENZYMATIC HYDROLYSIS

A 22 central composite design was used to optimize the enzymatic hydrolysis conditions. The statistical software Minitab (version 6.0, Minitab Inc., PA, USA) was used to generate the design of experiment, perform the statistical analysis, and develop the regression model. An enzyme loading of 5 to 40 mg EP/g glucan and SC 5 to 20 % were chosen as the independent variables. Fourteen experimental runs with six center points were conducted to evaluate the effect of enzyme loading and SC on glucose yield after enzymatic hydrolysis. Table 2 shows the design matrix and glucose yield response of the experiment. Data obtained from experimental runs were then analyzed according to the response surface regression and then fit to a second-order polynomial equation (Eq. (2)).

$$Y_i = \beta_0 + \sum_{i=2}^{n} \beta_i X_i + \sum_{i=2}^{n} \beta_{ii} X_i^2 + \sum_{i=1}^{n-1} \sum_{j=i+1}^{n} \beta_{ij} X_i X_j \quad (2)$$

where Y i is the response value, β 0 is the constant coefficient, β i is the linear coefficients, β ij (i and j) is the interaction coefficients, β ii is the quadratic coefficients, and χ i is the variables. Analysis of variance (ANOVA) was used to estimate the significance of single factors, factors' square, and interaction of factors. The relationship between dependent and independent variables was further determined by the response surface plot.

TABLE 2: Design of experiments for determining the optimal conditions of enzymatic hydrolysis using response surface methodology with predicted and experimental response.

Run order	Variables		Coded levels		Experimental glucose yield (%)	Predicted glucose yield (%)
	TS (%)	Protein (mg/g glucan)	$\chi 1$	$\chi 2$		
	$\chi 1$	$\chi 2$				
EH-1	5.00	5.00	−1	−1	58.13	57.54
EH-2	20.00	5.00	1	−1	62.14	61.47
EH-3	5.00	40.00	−1	1	77.12	78.90
EH-4	20.00	40.00	1	1	100.00	101.70
EH-5	12.50	22.50	0	0	87.21	90.28
EH-6	12.50	22.50	0	0	89.11	90.28
EH-7	12.50	22.50	0	0	90.07	90.28
EH-8	5.00	22.50	−1	0	84.41	83.22
EH-9	20.00	22.50	1	0	97.62	96.59
EH-10	12.50	5.00	0	−1	60.78	62.04
EH-11	12.50	40.00	0	1	96.31	92.83
EH-12	12.50	22.50	0	0	89.75	88.13
EH-13	12.50	22.50	0	0	91.68	88.13
EH-14	12.50	22.50	0	0	85.18	88.13

3.2.6 ANALYTICAL METHODS

A portion of the slurries obtained after WEx pretreatment and enzymatic hydrolysis were separated into solid and liquid fractions using a bench-top centrifuge (Eppendorf, Model 5804, 8,000 rpm, 10 min). The liquid portion was then decanted and filtered through a 0.45 μm syringe filter, and solids were washed and dried at 30 °C for 48 h. Free sugars in the liquid portion were analyzed using a Bio-Rad (Hercules, CA, USA) Aminex HPX-87P column with RI detector, operating at 83 °C with Milli-Q water (Barnstead Nanopure, USA) as an eluent with a flow rate of 1.0 mL/min. The composition of the solid fraction was determined as per established NREL method [14].

3.2.7 SCANNING ELECTRON MICROSCOPY

The morphological features of raw, pretreated, and enzymatically treated corn stover were observed by scanning electron microscopy (SEM) (S-570, Hitachi Ltd., Japan) at an acceleration voltage of 10 kV. SEM was conducted under high vacuum to improve the conductivity of the samples and the resolution of the SEM micrographs.

3.3 RESULTS AND DISCUSSIONS

3.3.1 RAW MATERIAL

The chemical composition of corn stover as determined using standard NREL protocols ("Analytical Methods") showed the following compositions: glucan 38.67 % (±0.12), xylan 25.21 % (±0.03), galactan 1.83 % (±0.03), arabinan 2.85 % (±0.05), mannan 0.38 % (±0.16), Klason lignin 17.52 % (±0.14), acid-soluble lignin 1.50 % (±0.01), ash 2.57 % (±0.08), and acetate 4.34 % (±0.01) on dry basis. The composition analysis was performed in duplicates, and numbers shown are the average and the standard error is shown within the parentheses. The composition of raw material was used to calculate the glucose yields (percent) in the experiments.

3.3.2 SELECTION OF PLATFORM WEx PRETREATMENT CONDITIONS

The amount of sugars released in the liquid phase after WEx pretreatment at 12 different conditions is shown in Fig. 2a. The amount of free sugars (glucose, xylose, and arabinose) released after WEx varied between 11.34 and 40.62 g/L. During the hydrolysis testing, the amount of free sugars released varied between 77.16 and 130.94 g/L (Fig. 2b). The amount of sugar degradation products (HMF and furfural) along with acetate (released from hemicellulose) generated during WEx varied between 3.79 and 12.79 g/L, whereas, during the enzymatic hydrolysis, their concentration varied between 6.50 and 15.49 g/L. The increase in HMF and furfural concentration during enzymatic hydrolysis is attributed to continued sugar degradation at a slower rate during enzyme hydrolysis. Interestingly, the pretreated sample CS-8 (170 °C, 20 min, 5.5 bar oxygen) released lower sugars during pretreatment (16.44 g/L) as compared to other pretreated samples which released maximum sugars during enzyme hydrolysis (130.94 g/L). These results are probably due to the fact that more oligomers were found in this sample (CS-8) due to a better opening of the structure (loosening of linkages between macromolecules in the cell wall). This is in agreement with previous studies by Ladisch et al. where they found that the optimal pretreatment conditions correlate with the oligosaccharides and not the monosaccharides released during pretreatment [15]. Sample CS-8 after pretreatment showed monomeric glucose (2.97 g/L), xylose (11.16 g/L), arabinose (2.31 g/L) and, after enzymatic hydrolysis, showed highest concentration of glucose monomers (102.61 g/L), xylose monomers (26.25 g/L), and arabinose (2.08 g/L), which gives amonomeric glucose yield of 81 % compared to the glucan concentration of the corn stover. The amount of degradation products in sample CS-8 after pretreatment was 3.99 g/L for HMF, 0.50 g/L for furfural, and 7.57 g/L for acetate. The amount of degradation products after enzymatic hydrolysis was 3.4 g/L for HMF, 0.46 g/L for furfural, and 9.6 g/L for acetate. The amount of lignin solubilized during pretreatment was 8 %, which remained unchanged during enzymatic hydrolysis. Sample CS-8 was, therefore, chosen as platform WEx pretreatment conditions for enzyme optimization study as it released highest sugars and minimal sugar degradation products.

3.3.3 ENZYME OPTIMIZATION

3.3.3.1 OPTIMIZATION OF ENZYMATIC HYDROLYSIS PARAMETERS

Full-factorial statistical model was developed by the statistical software Minitab 16. Table 2 shows coded and uncoded variables along with the experimental and model-predicted response. Low, midpoint, and high levels were coded as −1, 0 and +1, respectively. The response of glucose yield af-

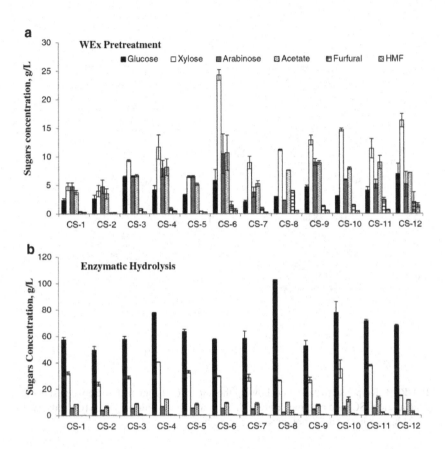

FIGURE 2: Sugars released after a WEx at 12 different pretreatment conditions and b enzymatic hydrolysis using 40 mg EP/g of glucan. The error bars in the plot show the standard deviation.

ter 72 h of enzymatic hydrolysis from all 14 runs as a function of cellulase loading (mg EP/g glucan) and SC was evaluated in CCD. The interaction between response and variables was evaluated by response surface after 12, 24, 48, and 72 h (Fig. 3). Each response surface plot confirmed the interactions existing between the variables and glucose yield. As shown in Table 2, the experimental glucose yield matched closely with the model-predicted glucose yield with a regression coefficient of 97.68 %.

Minitab 16 was used to conduct ANOVA to evaluate individual and interactive effects of variables. Significance of coefficients of the model

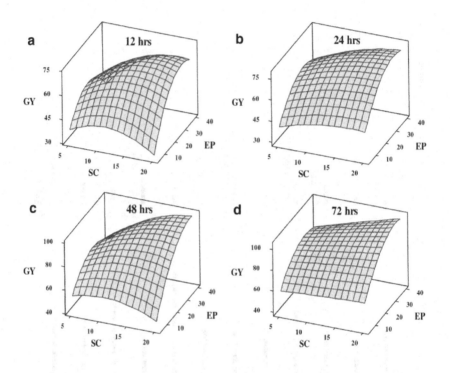

FIGURE 3: Response surface plot showing the impact of the interaction between substrate concentration (SC) and enzyme protein (EP) on monomeric glucose yields (GY) after **a** 12 h, **b** 24 h, **c** 48 h, and **d** 72 h of enzymatic hydrolysis.

was analyzed. The results of the second-order response surface models for the glucose released during hydrolysis are presented in the form of ANOVA in Table 3. Equation (3) describes the second-order polynomial model to evaluate the relationship between variables and glucose yield of enzymatic hydrolysis.

$$Y = 46.0929 + 0.2501_{X1} + 2.3183_{X2} - 0.0067^2_{X1} - 0.0420^2_{X2} + 0.0359_{X1X2} \quad (3)$$

TABLE 3: ANOVA table for glucose yield.

Source of variation	df	Sum of squares	Mean square	F ratio	p value	Coefficient
Regression	5	2,309.63	461.93	55.94	0.000	46.0929
Linear	2	1,690.35	845.17	102.34	0.000	–
SC (%)	1	268.00	268.00	32.45	0.000	0.2501
EP (mg)	1	1,422.34	1,422.34	172.24	0.000	2.318
Square	2	530.27	265.13	32.11	0.000	–
SC (%) × SC (%)	1	0.39	0.39	0.05	0.835	−0.0067
EP (mg) × EP (mg)	1	450.15	450.15	54.51	0.000	−0.0420
Interaction	1	89.02	89.02	24.05	0.013	–
SC(%) × EP (mg)	1	89.02	89.02	24.05	0.013	0.0359
Residual error	7	57.81	8.26	–	–	–
Lack of fit	3	31.28	10.43	1.57	0.33	–
Pure error	4	26.52	6.63	–	–	–
Total	13	2,547.84	–	–	–	–

$R^2 = 97.68$ %, R^2 (predicted) $= 85.94$ %, R^2 (adjusted) $= 95.69$ %

In the equation on page 32 (Eq. (3)), Y is the glucose yield after 72 h of hydrolysis, $\chi 1$ is SC (percent), and $\chi 2$ is enzyme loading (milligrams EP per gram glucan). The statistical significance of the model was evaluated by F test for ANOVA, which confirmed that the regression was statistically significant. Temperature and pH were excluded from the optimization model because these factors had already been optimized as 50 °C and 5, respectively (www.bioenergy.novozymes.com).

Statistical effect of SC and enzyme loading on glucose yield is shown in Table 3. As shown in Table 3, the coefficient of regression (R^2) was calculated as 97.68 %, which indicated that the model is suitable to adequately represent the real relationships among the selected reaction variables. Highly adjusted R^2 (95.69 %) also showed the suitability of the model. Linear and squared terms of enzyme loading exerted a positive effect on the yield of glucose during enzymatic hydrolysis as the p values were <0.05. Linear SC has been found to have a significant effect; however, there was no direct effect of squared SC. This can be explained by the heterogeneity of the pretreated sample. Interaction of both enzyme loading and SC has a significant effect ($p < 0.05$), which indicates cumulative effect of both factors on enzymatic hydrolysis.

The objective of the optimization was to achieve a glucose yield of at least 80 %. The SC and enzyme loading combination at which the highest desirability (0.959) was obtained was chosen as the optimal conditions for the enzymatic hydrolysis with the predicted value of the response, 85.94 %. The optimal conditions for the enzymatic hydrolysis from the statistical analysis were found to be 20 % SC and 15 mg EP/g glucan (about 5 FPU), which gave a glucose yield of 84 % that is in agreement with the predicted response. Thus, we concluded that the model was valid and reliable in predicting the experimental results.

3.3.3.2 INFLUENCE OF SC AND ENZYME LOADING DURING ENZYMATIC HYDROLYSIS

The significance of the quadratic coefficients of enzyme loading and SC for corn stover indicates a positive correlation between enzyme loading, SC, and hydrolysis yield. No noticeable mass transfer limitation was observed (due to high viscosity of the slurry) in contrast to previous studies [16, 17] as evident by high glucose yield. Optimal process conditions for enzymatic hydrolysis to high glucose included both high SC and low enzyme loading. This is a remarkable improvement over previous studies done with pretreated corn stover either at low SC (<10 %) or high enzyme loading (>10 FPU, equivalent to 30 mg EP/g glucan) to achieve higher cellulose conversion [5, 6, 18, 19]. By increasing the SC, the glucose yield improved significantly. This lowers the cost of fermentation and eliminates the need for concentration of the hydrolysate before fermentation, which would add to the overall cost.

3.3.4 TEST AT SCALE-UP CONDITIONS OF ENZYMATIC HYDROLYSIS

Under the optimal conditions (20 % SC, 15 mg EP/g glucan, pH 5, and 50°C), the model predicted a glucose yield of 84.8 %. The actual experiment yield was 84 %, which was in close agreement with the model's prediction, confirming the validity of the RSM model. A comparative study in 100-L pilot scale was performed with the optimized hydrolysis condition (20 % SC, 15 mg EP/g glucan) and compared to the hydrolysis condition at which the maximum yield was obtained (20 % SC, 40 mg EP/g glucan). Glucose yield of 84 and 99 % was attained at 15 and 40 mg EP/g glucan, respectively.

3.3.4.1 SCANNING ELECTRON MICROSCOPY

SEM was performed to understand the morphological changes for the corn stover samples when hydrolyzed at the determined enzyme loading (15 mg EP/g glucan) versus the highest enzyme loading (40 mg EP/g glucan). Figure 4 shows SEM micrographs to elucidate the changes in the fibrillar structure of raw corn stover and after treatment (WEx and enzymatic hydrolysis). Corn stover (Fig. 4a) exhibited compact and highly ordered fibril structure covered with a wax layer, binding materials, and other deposits (common to herbaceous lignocellulosic biomass). Figure 4b showed damages in the fibrillar structure as a result of WEx pretreatment. Degradation caused by the harsh pretreatment conditions was demonstrated by the appearance of exposed smooth cellulose surface, peeling, reticular areas, and interfibrillar splitting in the direction of the fiber axis. Despite these major changes after WEx, some conducting vessels and pits still remained intact, demonstrating the need for enzyme digestion. Panels c and d of Fig. 4 show the SEM images after enzymatic hydrolysis conducted

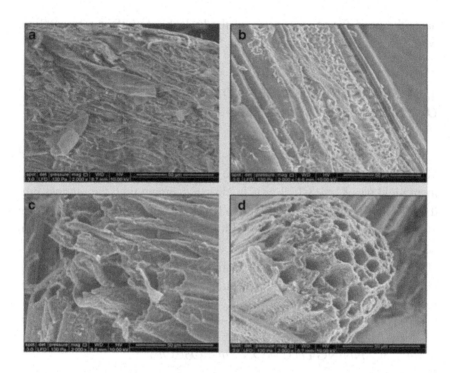

FIGURE 4: Scanning electron micrograph of corn stover: **a** control, raw corn stover; **b** wet-oxidized corn stover; **c** enzymatically treated corn stover using 15 mg EP at 20 % SC; and **d** enzymatically treated corn stover using 40 mg EP at 20 % SC. Scale: **a–d**=50 μm.

at enzyme loading of 15 and 40 mg EP/g glucan, respectively. Enzymatic hydrolysis caused the fibrous surface to be altered and enhanced the porosity induced by the WEx pretreatment. In both panels c and d of Fig. 4, degradation of fiber surface was significant, but the extent of degradation varied with the amount of enzyme protein added (and cellulase accessibility to fibers). With the determined optimal enzymes loading of 15 mg EP/g glucan (Fig. 4c), longitudinal disintegration of fibers and formation of fissures on the surface were observed and surface erosion was unevenly distributed, whereas with the increase in enzyme loading (Fig. 4d), uniform surface corrosion was observed. These observations lead to the conclusion that morphological differences between the corn stover samples

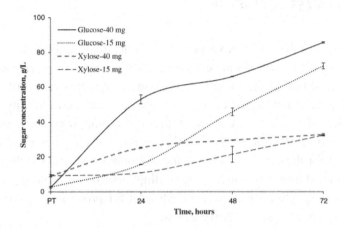

FIGURE 5: Glucose and xylose concentration (grams per liter) at 20 % substrate concentration and 40 and 15 mg protein at pilot scale.

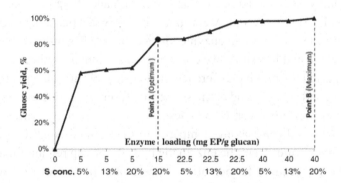

FIGURE 6: Effect of enzyme loading and substrate concentration on glucose yield.

hydrolyzed with 15 and 40 mg EP/g glucan were noticeable but relatively minor compared to the fact that 2.7 times more enzyme was added in the last case (Fig. 5).

3.3.5 PROCESS ECONOMICS

Figure 6 shows the effect of varying enzyme loadings and SC on glucose yields. With the increase in the enzyme loading and SC, high glucose yield was achieved but, at the same time, enzyme cost increased significantly due to the need for higher enzyme dosage. Due to unavailability of an exact price of Cellic® CTec2, we found it more appropriate to use the relative cost savings derived from the quantity of enzymes used at the sweet spot for the economics modeling. Amount of enzymes (kilograms) was calculated from enzyme loading (milligrams enzyme protein per gram glucan) at respective substrate concentration for processing 1,000 kg of biomass slurry as basis as shown below.

Using 210 mg EP/mL of enzyme solution, enzyme density as 1.2 g/mL, SC 20 %, glucan 38.67 %, and enzyme loading at points A and B as 15 and 40 mg EP/g glucan, respectively, the amount of enzyme calculated at points A and B was 6.51 and 17.37 kg, respectively, to process 1,000 kg of biomass slurry. Glucose yields at points A and B were 84 and 99 %, respectively. At point A, our sweet spot, we observed a plateau, which suggests that further increase in enzyme loading and SC has little effect on the glucose yield, and therefore, we considered that point as optimal.

Comparing the relative difference in glucose yield and enzyme amount at points A and B, we found at the sweet spot (point A) a glucose yield that was 16 % lower; however, 62.5 % less enzymes were used, which would represent a significant saving compared to the optimal condition. Moreover, some enzymes from the enzymatic hydrolysis step remain functional during the subsequent fermentation process, and these enzymes will be capable of processing at least parts of the 16 % of unconverted cellulose remaining after enzymatic hydrolysis into glucose. These findings agree with the past studies done by Shen and Agblevor [20], who suggested that the enzyme cost for complete hydrolysis (100 % conversion of cellulose to glucose) is prohibitive in achieving the economically viable commercialization of ethanol production.

3.4 CONCLUSION

WEx pretreatment and enzymatic hydrolysis at high SC were found to produce high glucose yield from corn stover even at low enzyme loadings. Experimental results showed that the most optimal and economically feasible condition for high yields of glucose (about 84 %) was obtained when corn stover was wet-exploded at 170 °C for 20 min with 5.5 bar oxygen and subsequently enzymatically hydrolyzed at 20 % SC and 15 mg EP/g glucan. Significant enzyme cost saving can be achieved by accepting a glucose yield of about 80 % compared to the full conversion of the glucan of corn stover.

REFERENCES

1. Banerjee G, Scott-Craig J, Walton J (2010) Improving enzymes for biomass conversion: a basic research perspective. Bioenergy Res 3:82–92
2. Lynd LR, Laser MS, Bransby D, Dale BE, Davison B, Hamilton R, Himmel M, Keller M, McMillan JD, Sheehan J, Wyman CE (2008) How biotech can transform biofuels. Nat Biotechnol 26:169–172
3. Kristensen J, Felby C, Jorgensen H (2009) Yield-determining factors in high-solids enzymatic hydrolysis of lignocellulose. Biotechnol Biofuels 2:11
4. Kaar WE, Holtzapple MT (2000) Using lime pretreatment to facilitate the enzymic hydrolysis of corn stover. Biomass Bioenergy 18:189–199
5. Kim S, Holtzapple MT (2005) Lime pretreatment and enzymatic hydrolysis of corn stover. Bioresour Technol 96:1994–2006
6. Kim TH, Lee YY (2005) Pretreatment of corn stover by soaking in aqueous ammonia. Appl Biochem Biotechnol 121:1119–1131
7. Bura R, Chandra R, Saddler J (2009) Influence of xylan on the enzymatic hydrolysis of steam-pretreated corn stover and hybrid poplar. Biotechnol Progr 25:315–322
8. Zhu JY, Zhu W, Obryan P, Dien B, Tian S, Gleisner R, Pan XJ (2010) Ethanol production from SPORL-pretreated lodgepole pine: preliminary evaluation of mass balance and process energy efficiency. Appl Microbiol Biotechnol 86:1355–1365
9. Yang J, Zhang XP, Yong QA, Yu SY (2011) Three-stage enzymatic hydrolysis of steam-exploded corn stover at high substrate concentration. Bioresour Technol 102:4905–4908

10. Ahring BK, Munck J (2009) Method for treating biomass and organic waste with the purpose of generating desired biologically based products. US Patent 0178671

11. Ahring BK, Jensen K, Nielsen P, Bjerre AB, Schmidt AS (1996) Pretreatment of wheat straw and conversion of xylose and xylan to ethanol by thermophilic anaerobic bacteria. Bioresour Technol 58:107–113

12. Adney B, Baker J (2008) Measurement of cellulase activities. National Renewable Energy Laboratory, Golden

13. Rana D, Rana V, Ahring BK (2012) Producing high sugar concentrations from loblolly pine using wet explosion pretreatment. Bioresour Technol 121:61–67

14. Sluiter A, Hames B, Ruiz R, Scarlata C, Sluiter J, Templeton D, Crocker D (2008) Determination of structural carbohydrates and lignin in biomass. National Renewable Energy Laboratory, Golden

15. Wyman CE, Dale BE, Elander RT, Holtzapple M, Ladisch MR, Lee YY, Mitchinson C, Saddler JN (2009) Comparative sugar recovery and fermentation data following pretreatment of poplar wood by leading technologies. Biotechnol Prog 25:333–339

16. Ingesson H, Zacchi G, Yang B, Esteghlalian AR, Saddler JN (2001) The effect of shaking regime on the rate and extent of enzymatic hydrolysis of cellulose. J Biotechnol 88:177–182

17. Wen ZY, Liao W, Chen SL (2004) Hydrolysis of animal manure lignocellulosics for reducing sugar production. Bioresour Technol 91:31–39

18. He X, Miao YL, Jiang XJ, Xu ZD, Ouyang PK (2010) Enhancing the enzymatic hydrolysis of corn stover by an integrated wet-milling and alkali pretreatment. Appl Biochem Biotechnol 162:2449–2457

19. Varga E, Szengyel Z, Reczey K (2002) Chemical pretreatments of corn stover for enhancing enzymatic digestibility. Appl Biochem Biotechnol 98:73–87

20. Shen J, Agblevor FA (2008) Optimization of enzyme loading and hydrolytic time in the hydrolysis of mixtures of cotton gin waste and recycled paper sludge for the maximum profit rate. Biochem Eng J 41:241–250

CHAPTER 4

Bioconversion of Lignocellulose: Inhibitors and Setoxification

LEIF J. JÖNSSON, BJÖRN ALRIKSSON
AND NILS-OLOF NILVEBRANT

4.1 REVIEW

4.1.1 BACKGROUND

Lignocellulose provides an abundant renewable resource for production of biofuels, chemicals, and polymers [1-3]. Biorefineries, in which lignocellulosic biomass is converted to various commodities, are likely to become increasingly important in future society as complement and alternative to the oil refineries of today. Commodities produced from renewable resources offer an alternative to products based on dwindling supplies of petroleum and permit a move towards improved energy security and decreased impact on the environment. Lignocellulosic feedstocks include residues from agriculture and forestry, energy crops, and residues from biorefineries and pulp mills. Lignocellulosic biomass can contribute significantly to the future global energy supply without competition with increasing food demand for existing arable land [4].

Liquid biofuels include bioalcohols, such as ethanol and butanol, and biodiesel. Ethanol is the most important liquid biofuel of today.

Bioconversion of Lignocellulose: Inhibitors and Setoxification. © 2013 Jonsson et al.; licensee BioMed Central Ltd. Biotechnology for Biofuels 2013, 6:16 (doi:10.1186/1754-6834-6-16).Creative Commons Attribution License (http://creativecommons.org/licenses/by/2.0).

Bioalcohols are manufactured in fermentation processes, in which microbial biocatalysts, yeasts or bacteria, convert sugars to alcohols. The ethanol that is used today is mainly manufactured from sugar or starch-based raw materials. However, very large-scale use of bioalcohols in the energy sector will require production from lignocellulosic feedstocks [1-5], which have the added benefit that they are not used for food. This review focuses on biocatalyst inhibitors formed during acidic thermochemical pretreatment of lignocellulosic feedstocks, and how conditioning of slurries and hydrolysates can be used to alleviate inhibition problems connected with hydrolytic enzymes and the yeast *Saccharomyces cerevisiae*.

4.1.2 LIGNOCELLULOSE AND PRETREATMENT OF LIGNOCELLULOSIC FEEDSTOCKS

Lignocellulosic feedstocks mainly consist of cellulose, hemicellulose, and lignin [6,7]. Cellulose is an unbranched homopolysaccharide consisting of D-glucopyranosyl units. Hemicelluloses are branched heteropolysaccharides consisting of both hexose and pentose sugar residues, which may also carry acetyl groups. The third main component, lignin, consists of phenylpropane units linked together by different types of interunit linkages of which ether bonds are the most common. Lignocellulose polysaccharides are hydrolyzed to provide the monosaccharides used by microbial biocatalysts in fermentation processes. The crystalline parts of the cellulose are more resistant to hydrolysis than are the amorphous parts. Compared to starch, the polysaccharides of lignocellulose are more resistant to hydrolysis. Furthermore, woody biomass is generally more resistant to degradation than other types of lignocellulose. Softwood is typically more difficult to hydrolyze than hardwood or agricultural residues [8-12].

Hydrolysis of cellulose can be catalyzed by using strong inorganic acids or hydrolytic enzymes, including cellulases [13,14]. Acid hydrolysis of cellulose requires severe conditions. Enzymatic hydrolysis is often considered as the most promising approach for the future [5]. Lignocellulosic biomass intended for production of liquid biofuels is typically pretreated in an acidic thermochemical process step to increase the susceptibility of the cellulose to enzymatic hydrolysis [5,9,12]. The pretreatment usually degrades the hemicellulose leading to the formation of products such as pentose and hexose sugars, sugar acids, aliphatic acids (primarily acetic acid, formic acid and levulinic acid), and furan aldehydes [5-hydroxymeth-ylfurfural (HMF) and furfural] (Figure 1). After hydrolysis of lignocellu-lose polysaccharides, lignin remains as a solid residue, although a minor part is degraded to phenolics and other aromatic compounds (Figure 1). Sugars derived from hemicelluloses will account for a substantial part of the total sugar and it is desirable that they are included in the subsequent fermentation step. The monosaccharides obtained through the hydrolysis process are then fermented by microbial catalysts to the desired product, most commonly ethanol produced with the yeast *S. cerevisiae*.

Hydrolysis and fermentation can be performed separately (separate hydrolysis and fermentation; SHF) or simultaneously (simultaneous

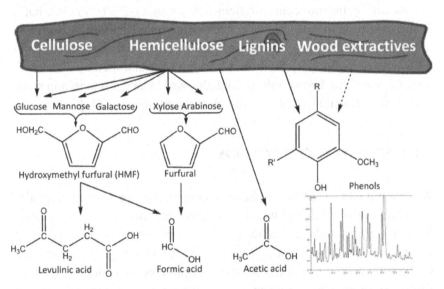

FIGURE 1: Formation of inhibitors. Scheme indicating main routes of formation of inhibitors. Furan aldehydes and aliphatic acids are carbohydrate degradation products, while lignin is the main source of phenolic compounds, as indicated by guaiacyl (4-hydroxy-3-methoxyphenyl) and syringyl (4-hydroxy-3,5-dimethoxyphenyl) moieties found in many phenolics. While the contents of furan aldehydes and aliphatic acids are relatively easy to determine, the quantification and identification of phenolic compounds remain challenging. The insert shows the variety of peaks representing phenolic compounds found in a hydrolysate of Norwegian spruce, as indicated by analysis using liquid chromatography-mass spectrometry (LC-MS)..

saccharification and fermentation; SSF). Consolidated bioprocessing (CBP) refers to a process in which the fermenting microorganism also contributes by producing cellulolytic enzymes [15].

4.1.3 INHIBITORS OF ENZYMATIC AND MICROBIAL BIOCATALYSTS

The generation of by-products from the pretreatment is strongly dependent on the feedstock and the pretreatment method. Substances that may act as inhibitors of microorganisms include phenolic compounds and other aromatics, aliphatic acids, furan aldehydes, inorganic ions, and bioalcohols or other fermentation products. Examples of inhibitory fermentation products are ethanol and butanol. As most microorganisms, S. cerevisiae is inhibited by butanol concentrations in the range 1-2% (v/v) [16], but it is able to withstand much higher concentrations of ethanol. In high-gravity alcoholic fermentations, *S. cerevisiae* produces ethanol concentrations of 17% (v/v) or higher [17]. Hydrolytic enzymes are inhibited by their products, i.e. sugars such as cellobiose and glucose [18], by fermentation products such as ethanol [19,20], and by phenolic compounds [21].

4.1.4 AROMATIC COMPOUNDS

A large number of different phenolic compounds are formed from lignin during acid-catalyzed hydrolysis or pretreatment of lignocellulose. Phenolic compounds and other aromatics are formed during pretreatment regardless of whether an acid catalyst is added to the reaction [22]. Carboxylic acids formed during the pretreatment will contribute to the formation of an acidic environment. Furthermore, some extractives are phenolic compounds [6,7]. Formation of phenolic compounds from sugars is another possibility [23], although the significance of this route remains to be investigated.

Different analytical techniques, primarily gas chromatography–mass spectrometry (GC-MS) and liquid chromatography-mass spectrometry (LC-MS), have been used to identify specific aromatic compounds in acidic hydrolysates from various kinds of lignocellulosic feedstocks, such as corn stover [24-26], oak [27], pine [26,28,29], poplar [24,30-32], spruce [33-35], sugarcane bagasse [22], switchgrass [24], and willow [36]. In addition, aromatic degradation products in hydrolysates produced by alkaline methods have been investigated [26,37]. The large number and the diversity of the aromatic compounds found in different lignocellulose hydrolysates (Figure 1) make identification and quantification of separate compounds complicated. Group analysis of phenolic compounds offers an alternative approach. GC-MS has been used to estimate the total amount of phenols in lignocellulose hydrolysates [33,36]. The total amount of phenols in a spruce wood hydrolysate was determined spectrophotometrically by using the Prussian Blue method [33]. Persson et al. [34] compared the Prussian Blue method with another spectrophotometric method, based on Folin-Ciocalteu's reagent, and found that the latter gave more reliable results with respect to analysis of phenolic compounds in the hydrolysate. A peroxidase-based biosensor was also tested, as an alternative to the spectrophotometric methods [34]. Furthermore, a method for group analysis of phenols by high-performance liquid chromatography (HPLC) has also been used [38]. Although the Folin-Ciocalteu method is the most convenient approach to analyze the total phenolic contents in lignocellulose hydrolysates, it should be avoided in experiments with redox reagents (such as reduced sulfur compunds including dithionite, dithiothreitol, and sulfite), in which the HPLC method serves as a better option [39]. It should also be noticed that phenol analysis using the Folin-Ciocalteu reagent is related to the Lowry method for determination of the total protein content [40] and that it is therefore sensitive to potential media components such as hydrolytic enzymes, cell extracts, and hydrolyzed protein.

The effects of phenolics and other aromatic compounds, which may inhibit both microbial growth and product yield, are very variable, and can be related to specific functional groups [30,41]. In many cases, the mechanism of toxicity has not been elucidated. One possible mechanism is that phenolics interfere with the cell membrane by influencing its function and changing its protein-to-lipid ratio [42]. *S. cerevisiae* can convert some inhibitory phenolics to less toxic compounds. For instance, coniferyl aldehyde is reduced to coniferyl alcohol and dihydroconiferyl alcohol [41].

The role of phenolic inhibitors has been investigated using enzymic catalysts that specifically affect phenolic compounds without changing the concentrations of other inhibitors, such as aliphatic acids and furan aldehydes [33,36,43-45]. Enzymes, such as laccases and peroxidases, oxidize phenols to radicals that undergo coupling to larger molecules that are less toxic to fermenting microbes such as yeast [36].

Phenolic compounds are also investigated with regard to inhibition of enzymatic hydrolysis of cellulose [21]. Experiments with phenols suggest that one way in which they affect proteins is by inducing precipitation [46].

4.1.5 ALIPHATIC ACIDS

Lignocellulose hydrolysates contain aliphatic acids, such as acetic acid, formic acid, and levulinic acid. Acetic acid is formed primarily by hydrolysis of acetyl groups of hemicellulose, while formic acid and levulinic acid arise as acid-catalyzed thermochemical degradation products from polysaccharides (Figure 1). Formic acid is a degradation product of furfural and HMF (5-hydroxymethylfurfural), while levulinic acid is formed by degradation of HMF [47]. The pKa value of formic acid (3.75) is considerably lower than those of acetic acid (4.76) and levulinic acid (4.64). The toxic effect on *S. cerevisiae* is attributed to the undissociated form and increases in the order acetic acid < levulinic acid < formic acid. Inhibition of yeast was found to be apparent at concentrations exceeding 100 mM [48]. However, lower concentrations than 100 mM gave higher ethanol yields than fermentations with no aliphatic acids included [48]. The contents of aliphatic acids in slurries and hydrolysates vary strongly depending on the feedstock and the severity of the pretreatment. Feedstocks with high content of acetylated xylan, typically agricultural residues and hardwood, give higher concentrations of aliphatic acids than softwood. The total content of aliphatic acids in softwood hydrolysates is often below 100 mM and consequently beneficial for the ethanol yield rather than harmful [48,49].

Undissociated acids enter the cell through diffusion over the cell membrane and then dissociate due to the neutral cytosolic pH [50]. The dissociation of the acid leads to a decrease in the intracellular pH, which may lead to cell death. Alternatively, it may lead to increased ethanol yield at the expense of biomass formation as a consequence of the cell's attempt to maintain a constant intracellular pH by pumping out protons through the plasma membrane ATPase [51-53].

A group of compounds that can be mentioned in this context are uncouplers, i.e. amphiphilic molecules that dissolve in the inner mitochondrial membrane of eukaryotic cells and that have the ability to transfer protons across the membrane. By disrupting the proton gradient over the inner mitochondrial membrane, they disconnect the linkage between the respiratory chain and the oxidative phosphorylation that regenerates ATP from ADP. This mechanism differs from that proposed for aliphatic acids like acetic acid, as it inhibits the regeneration of ATP in mitochondria rather than stimulate the consumption of ATP at the plasma membrane. Some aromatic carboxylic acids may act as uncouplers, as has been shown in experiments with plant cells and salicylic acid [54], a compound that is also found in lignocellulose hydrolysates [26,36]. Another aromatic carboxylic acid, p-hydroxybenzoic acid, which is common in lignocellulose hydrolysates, did not exhibit the uncoupling effect observed for salicylic acid [54].

4.1.6 FURAN ALDEHYDES

The furan aldehydes furfural and HMF, which also are commonly found in lignocellulose hydrolysates, are formed by dehydration of pentose and hexose sugars, respectively (Figure 1). Furfural and HMF inhibit the growth of yeast and decrease ethanol yield and productivity [48,55,56]. Under anaerobic conditions, *S. cerevisiae* can convert furfural to furfuryl alcohol [57,58] and HMF to 2,5-bis-hydroxymethylfuran [59]. Reduction of furfural has been linked to the co-factor NADH, while reduction of HMF has been found to be associated with consumption of NADPH [60]. A moderate addition of furfural to the growth medium was found to lead to increased ethanol yields for recombinant xylose-utilizing S. cerevisiae transformants [60]. This can be explained by the reduction of furfural to furfuryl alcohol, which will lead to a decreased formation of the undesirable by-product xylitol and an increased formation of ethanol. Model fermentations with furan aldehydes added to the medium suggest that yeast can tolerate quite high concentrations of furan aldehydes [48,61]. Martinez et al. [62] noticed that it took an addition of three times the original concentrations of the furan aldehydes to restore the inhibition of E. coli by a detoxified bagasse hydrolysate. These observations suggest that the inhibition might be due to other inhibitors present in the hydrolysate, other yet unidentified compounds, or perhaps to synergistic effects involving furan aldehydes. The capability of the microorganism to reduce furan aldehydes to the less toxic corresponding alcohols during fermentation in a bioreactor is sometimes referred to as in-situ detoxification [63]. The concept of biological in-situ detoxification is based on the presumption that it is the mere presence of the inhibitory substance that is the problem, rather than its bioconversion.

4.1.7 INORGANIC COMPOUNDS

Inorganic ions that are present in lignocellulose hydrolysates originate from the lignocellulosic feedstocks, from chemicals added during pretreatment, conditioning and hydrolysis, and possibly from process equipment. The addition of salts results in a higher osmotic pressure, which may result in inhibitory effects [64,65]. At moderate concentrations, there is a possibility that inorganic ions enhance ethanol production in a similar way as moderate concentrations of aliphatic acids do. The proposed mechanism is increased demand of ATP due to increased transport over the plasma membrane. Extra ATP is acquired by an increased ethanol production at the expense of biomass formation.

    S. cerevisiae is relatively salt tolerant compared to other yeasts, such as Schizosaccharomyces pombe and Scheffersomyces (Pichia) stipitis, but less tolerant than several Candida species [64]. In glucose-based medium, S. cerevisiae is capable to grow in a 1.5 M solution of sodium chloride. However, a more important factor than the absolute concentration of sodium is the intracellular ratio of $Na+/K+$, which preferably should be kept low. Maiorella et al. [66] investigated the effects of different salts on S. cerevisiae and found that the inhibition decreased in the following order: $CaCl_2$, $(NH4)_2SO_4$ > NaCl, NH_4Cl > KH_2PO_4 > $MgCl_2$ > $MgSO_4$ > KCl.

4.1.8 OTHER INHIBITORY EFFECTS

Ethanol generated during fermentation inhibits viability, growth, glucose transport systems, and proton fluxes of *S. cerevisiae*. The yeast plasma membrane is affected with respect to permeability, organization, and lipid composition [67]. However, the ethanologenic microbes *S. cerevisiae* and *Zymomonas mobilis* can tolerate ethanol concentrations up to 18 and 12%, respectively [68]. The engineering of microbes for improved resistance to bioalcohols and other biofuels has recently been reviewed [16].

Potential synergistic effects of inhibitors have been studied in experiments with yeast and bacteria [69-71]. The results of these studies indicate synergistic effects of combinations of acids and furan aldehydes, as well as of combinations of different phenolics.

4.1.9 STRATEGIES TO COUNTERACT INHIBITION PROBLEMS

Several alternative measures can be taken to avoid problems caused by inhibitors. The concentrations of inhibitors and sugars in hydrolysates depend on the feedstock as well as on the conditions during pretreatment and hydrolysis [9,48]. Therefore, one possibility is to select less recalcitrant feedstocks and to utilize mild pretreatment conditions. However, it is desirable to utilize different varieties of lignocellulose if production of commodities from renewables should make a major impact on the market for fuels, chemicals, and materials. Furthermore, production of bulk chemicals is yield dependent, which implies that it is not reasonable to accept a poor sugar yield, and consequently a poor overall product yield, due to the use of insufficient pretreatment conditions.

It is also possible to design the fermentation process to avoid problems with inhibition, for example by using SSF to avoid inhibition of cellulolytic enzymes by sugars, or by using fed-batch or continuous cultivation rather than batch processes [72]. High yield and productivity, high product titer, and possibilities to recirculate process water are, however, important aspects of the chosen design. Ethanol production from diluted hydrolysates with low sugar content is associated with a high operating cost due to a more expensive distillation process [68].

There is a variety of different chemical, biological and physical methods that can be used to detoxify slurries and hydrolysates [33,73,74]. Approaches that have been studied include overliming and treatments with other chemicals, liquid-liquid extraction, liquid–solid extraction, heating and evaporation, and treatments with microbial and enzymatic biocatalysts (Table 1). Comparisons of different methods for detoxification, or conditioning, indicate that they differ significantly with respect to effects on hydrolysate chemistry and fermentability [33,75]. A common objection against detoxification is based on the assumption that it would require a separate process step.

There are a number of strategies that concern the fermenting microorganism. The use of large inocula decreases inhibition problems [55,73,75].

TABLE 1: Techniques for detoxification of lignocellulose hydrolysates and slurries.

Technique	Procedure	Example[a]
Chemical additives	Alkali [such as $Ca(OH)_2$, NaOH, NH_4OH]	[76,77]
	Reducing agents [such as dithionite, dithiothreitol, sulfite]	[39]
Enzymatic treatment	Laccase	[36,45]
	Peroxidase	[36]
Heating and vaporization	Evaporation	[33]
	Heat treatment	[78]
Liquid-liquid extraction	Ethyl acetate	[24,75]
	Supercritical fluid extraction [such as supercritical CO_2]	[34]
	Trialkylamine	[79]
Liquid–solid extraction	Activated carbon	[80]
	Ion exchange	[38,81]
	Lignin	[82]
Microbial treatment	*Coniochaeta ligniaria*	[83,84]
	Trichoderma reesei	[33,85]
	Ureibacillus thermosphaericus	[86]

[a]The table includes one or two examples of each procedure (references are not exhaustive). Dilution, washing of solid fractions, and techniques based on the fermenting microbe are not included.

However, the use of large inocula is considered to be a less attractive solution in an industrial context [87]. Using a large inoculum would be a possibility if the microorganism can be recirculated and reused at a reasonable cost. However, if the used fermentation broth contains a lot of solids, the separation of the microorganism could become a tedious task. This is the case in SSF processes, and as a consequence the use of fresh inocula is considered instead of recycling the microorganism [88].

Other possibilities that target the microorganism include selection of microbial species and strains that exhibit resistance to inhibitors. Adaptation of the microorganism to an inhibiting environment, possibly after inducing variation by mutagenesis, serves as an alternative option. Furthermore, genetic engineering can be employed to obtain transformed hyperresistant microbes. *S. cerevisiae* has been engineered for increased resistance to fermentation inhibitors by overexpression of enzymes conferring improved resistance to phenolics [89,90], furan aldehydes [91,92], and aliphatic acids [93,94]. Furthermore, overexpression of a transcription factor, Yap1 [95], and of multidrug-resistance proteins [95] has also generated hyperresistant *S. cerevisiae* transformants. In some of these cases, hyperresistance to lignocellulose hydrolysates has also been demonstrated [89,90,95].

Most of the studies on inhibition have had focus on the fermenting microorganism, while strategies that decrease inhibition of enzymes so far have received relatively little attention. Since most enzymatic hydrolysis processes involve mixtures of a pretreatment liquid and a solid cellulosic material, there are good reasons to take enzyme inhibition into account. Chemical detoxification, a powerful strategy to deal with inhibitor problems which also addresses enzyme inhibition, will be considered in more detail below.

4.1.10 CHEMICAL TREATMENT

Although methods such as liquid-liquid extraction, ion exchange, and treatment with biocatalysts remain frequently studied options for detoxifying hydrolysates or slurries, the focus of this section will be detoxification by addition of alkali or other chemical agents. In comparisons of

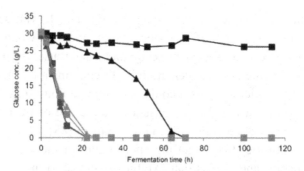

FIGURE 2: Effects of genetic engineering for hyperresistance and chemical detoxification through alkaline treatment. Ethanol production by *S. cerevisiae* (control transformant and transformant overexpressing Yap1 [95]): in spruce hydrolysate medium (black triangle, Yap1 transformant; black square, Control transformant), in alkali-detoxified spruce hydrolysate (dark gray triangle, Yap1 transformant; dark gray square, Control transformant), and in inhibitor-free medium (light gray triangle, Yap1 transformant; light gray square, Control transformant)...

detoxification methods, treatment with calcium hydroxide (overliming) has emerged as one of the most efficient methods [33,75]. In many cases, overliming also seems to be the most economical choice [78]. Although biotechnical methods (reviewed in [74,96]) are very promising in a longer perspective, they are seldom compared to conventional methods, such as alkaline detoxification. A comparison between the performance of a hyperresistant S. cerevisiae transformant overexpressing Yap1 [95] and the effect of alkaline detoxification is shown in Figure 2. The result indicates that both approaches have a very clear positive impact, but only the fermentation after alkaline detoxification reaches a similar level as that of the reference fermentation.

Overliming of hydrolysates produced by pretreatment of lignocellulose with sulfuric acid results in the precipitation of calcium sulfate (gypsum) [76,97]. This keeps the concentration of soluble salts at a low level, which is favorable for the fermentation process [76,97]. However, treatment of hydrolysates with other types of alkali, such as ammonium hydroxide, can result in a fermentability that is equal to or even better than that of hydrolysates treated with overliming [76].

Although the mechanism of overliming is still not completely elucidated, considerable progress has been made. Van Zyl et al. [98] suggested
that the detoxification effect of overliming was due to precipitation of
toxic substances. Persson et al. [35] collected and analyzed precipitated
material as well as the chemical composition of alkali-treated hydrolysates
and concluded that the detoxification effect was due to chemical conversion rather than to removal of precipitated inhibitors. Furthermore, a comparison of different types of alkali for treatment of hydrolysates showed
that it was possible to obtain an excellent ethanol yield (better than in a
reference fermentation with similar sugar content but without inhibitors)
after treatment with sodium hydroxide [77]. Since the treatment with sodium hydroxide did not give rise to any precipitate, this finding confirmed
the conclusions drawn regarding the effects of alkaline treatment [35].

A problem associated with alkali detoxification is that not only inhibitors are affected by the treatment, but also the sugars, which could lead to
reduced ethanol yields (Table 2). Nilvebrant et al. [99] studied the effects
of treatment time, temperature, and pH during alkali treatment of a spruce
hydrolysate. During treatment with alkali, xylose was slightly more easily
degraded than the other monosaccharides. Using similar conditions (time
period, pH, and temperature), the effect of calcium hydroxide was larger
than that of sodium hydroxide. More extensive sugar degradation during
alkaline treatment by overliming can be attributed to the stabilisation of
reactive enolate intermediates by calcium ions (Figure 3). The examples in
Table 2 indicate that too harsh conditions result in extensive sugar degradation, which also has an adverse effect on ethanol production. However,
it is also evident that a considerable improvement of the fermentability can
be gained with a very small loss of sugar (about 1%) (Table 2) indicating
that sugar loss is not always a valid objection to alkaline detoxification.

Ethanol production is often reported as the overall ethanol yield (OEY,
i.e. the yield calculated on the sugar content of the hydrolysate prior to
detoxification and given in percent of the maximum theoretical yield)
(Table 2). However, OEY does not take the relative fermentation improvement and the fermentation rate into account. A high OEY can be achieved
after an intolerably long fermentation time. Since it is difficult to evaluate the significance of the improvement in fermentability without having a reference fermentation to relate it to, it is highly recommended that

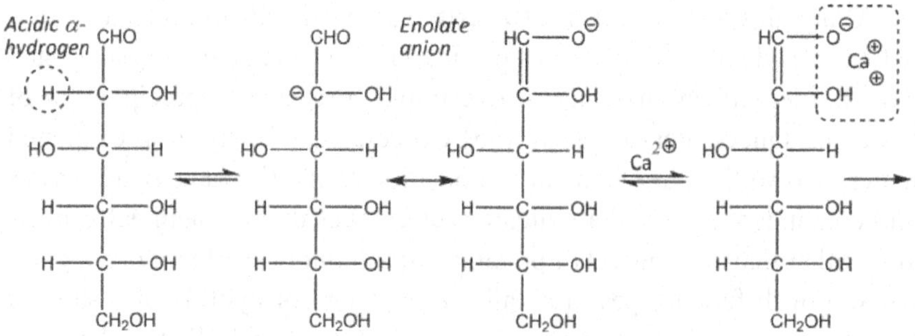

FIGURE 3: Monosaccharide degradation in alkali. Initial phase of degradation of glucose during alkaline treatment. Calcium ions stabilize the reactive enol intermediate, which in turn is degraded to HMF, and further to formic and levulinic acids.

reference fermentations without inhibitors should be included in detoxification studies. One possibility is to evaluate the treatment on basis of the balanced ethanol yield (BEY) (Table 2) [76]. BEY is the amount of ethanol produced divided by the total amount of fermentable sugars present in the hydrolysate prior to the detoxification given as percent of a reference fermentation of a sugar solution without inhibitors.

A new development in chemical detoxification is the possibility to perform the treatment in situ in the bioreactor by using reducing agents, such as sulfur oxyanions or sulfhydryl reagents [39]. Reducing agents eliminate the need for an extra process step for detoxification. Furthermore, treatment with reducing agents also decreases problems with inhibition of enzymatic hydrolysis [102]. The mechanism behind treatment with sulfur oxyanions such as bisulfite and dithionite was studied by Cavka et al. [61], who found that the effect was due to sulfonation of inhibitors, which rendered them unreactive and highly hydrophilic. The substances that are sulfonated by sulfur oxyanions include phenolics [61], which is noteworthy considering indications that phenolics play a role in the inhibition of enzymatic saccharification of cellulose [21,46].

TABLE 2: Effects of alkaline treatment on monosaccharides and ethanol production.

System studied	Detoxification conditions	Improvement in fermentability	Effect on inhibitors and sugar	Reference
Spruce hydroly-sate. S. cerevisiae	Ca(OH)$_2$ pH 10, 1 h	BEY[a]= 98%	Furan aldehydes, decrease: ~21%	[33]
		BEY[a] (untreated[b])= 71%	Phenols, decrease: ~19%	
		BEY[a] (reference[c])= 100%	Sugar, decrease: ~4%[d]	
	NaOH pH 10, 1 h	BEY[a]= 94%	Furan aldehydes, decrease: ~18%	
		BEY[a] (untreated[b])= 71%	Phenols, decrease: ~18%	
		BEY[a] (reference[c])= 100%	Sugar, decrease: ~4%[d]	
Bagasse hydroly-sate. E. coli	Ca(OH)$_2$ pH 9, 60°C, 0.5 h	Q (24 h)[e]: ~1.3 g/Lh	Furan aldehydes, decrease: ~69%	[97]
		No reference fermentation	Phenols, decrease: ~35%	
			Sugar, decrease: ~15%f	
	Ca(OH)$_2$ pH 10, 60°C, 0.5 h	Q(24 h)[e]: ~ 1.0 g/Lh	Sugar, decrease: ~33%f	
		No reference fermentation		
Spruce hydroly-sate. S. cerevisiae	Ca(OH)$_2$ pH 12, 60°C, 170 h	Q(24 h)[e]: ~ 0.3 g/Lh	Furan aldehydes, decrease: ~100%	[100]
		No reference fermentation	Phenols, increase: ~150%	
			Sugar, decrease: ~68%[g]	
	Ca(OH)$_2$ pH 11, 25°C, 20 h	Q(24 h)[e]: ~ 0 g/Lh	Furan aldehydes, decrease: ~77%	
		Q(48 h)[e]: ~ 0.3 g/Lh	Phenols, decrease: ~8%	
		No reference fermentation	Sugar, decrease: <5%[g]	
Bagasse hydroly-sate. S. cerevisiae	Ca(OH)$_2$ pH 10, 1 h	BEY[a]= 92%	Furan aldehydes, decrease: >25%	[43]
		BEY[a] (untreated[b])= 68%	Phenols, decrease: ~17%	
		BEY[a] (reference[c])= 100%	Sugar, decrease: ~1%f	

TABLE 2: CONTINUED.

Spruce hydroly-sate. S. cerevisiae	Ca(OH)$_2$ pH 11, 30°C, 3 h	BEY[a]= 120%	Furan aldehydes, decrease: ~59%	[49]
		BEY[a] (untreated[b])= 5%	Phenols, decrease: ~22%	
		BEYa (reference[c])= 100%	Sugar, decrease: ~14%[h]	
Spruce hydroly-sate. S. cerevisiae	NH$_4$OH pH 10, 22°C, 3 h	BEY[a] =110%	Not determined	[76]
		BEY[a] (untreated[b])= 10%		
		BEY[a] (reference[c])= 100%		
Spruce hydroly-sate. S. cerevisiae	NaOH pH 9, 55°C, 3 h	BEY[a] = 111%	Furan aldehydes, decrease: ~33%	[77]
		BEY[a] (untreated[b]) = 6%	Phenols, decrease: ~12%	
		BEY[a] (reference[c]) = 100%	Sugar, decrease: ~9%[d]	
	NH$_4$OH pH 9, 55°C, 3 h	BEY[a] =120%	Furan aldehydes, decrease: ~33%	
		BEY[a] (untreated[b]) = 7%	Phenols, decrease: ~13%	
		BEY[a] (reference[c]) = 100%	Sugar, decrease: ~7%[d]	
Corn stover hydrolysate. Z. mobilis	Ca(OH)$_2$ pH 9, 50°C, 0.5 h	No reference fermentation. OEY[i] = 62%	Sugar, decrease: ~7%[j]	[101]
	Ca(OH)$_2$ pH 10, 50°C, 0.5 h	No reference fermentation. OEY[i] = 70%.	Sugar, decrease: ~13%[j]	
	Ca(OH)$_2$ pH 11, 50°C, 0.5 h	No reference fermentation. OEY[i] = 59%	Sugar, decrease: ~29%[j]	

[a] Balanced ethanol yield given in percent of a reference fermentation of a sugar solution. [b] Untreated hydrolysate. [c] Reference sugar solution. [d] Glucose, xylose, arabinose, galactose mannose, and cellobiose. [e] Ethanol productivity. [f] Glucose, xylose, arabinose, galactose, and mannose. [g] Glucose, xylose, galactose, and mannose. [h] Glucose and mannose. [i] Overall ethanol yield, yield calculated on sugars present prior to detoxification, given in percent of the theoretical yield. [j] Glucose, xylose, and arabinose.

4.2 CONCLUSIONS

Acid-catalyzed thermochemical pretreatment of lignocellulosic feedstocks has several advantages: it is a simple and inexpensive approach for pretreatment that efficiently improves the susceptibility to cellulolytic enzymes, even for more recalcitrant types of lignocellulose. A drawback is the formation of by-products that inhibit enzymes and microorganisms in subsequent biocatalytic conversion steps. However, rapid progress in several areas, such as conditioning or detoxification of slurries and hydrolysates, fermentation technology, and microbial resistance to inhibitors, makes acid pretreatment into a highly competitive future alternative in the bioconversion of lignocellulosic feedstocks. Management of inhibition problems is likely to become more important in a development that favors flexibility with respect to feedstocks, processes based on high dry-matter content and high product concentrations, and recirculation of process water.

REFERENCES

1.   Ragauskas AJ, Williams CK, Davison BH, Britovsek G, Cairney J, Eckert CA, Frederick WJ Jr, Hallett JP, Leak DJ, Liotta CL, Mielenz JR, Murphy R, Templer R, Tschaplinski T: The path forward for biofuels and biomaterials. Science 2006, 311:484-489.
2.   Lynd LR, Laser MS, Bransby D, Dale BE, Davison B, Hamilton R, Himmel M, Keller M, McMillan JD, Sheehan J, Wyman CE: How biotech can transform biofuels. Nat Biotechnol 2008, 26:169-172.
3.   Sims REH, Mabee W, Saddler JN, Taylor M: An overview of second generation biofuel technologies. Bioresour Technol 2010, 101:1570-1580.
4.   Metzger JO, Hüttermann A: Sustainable global energy supply based on lignocellulosic biomass from afforestation of degraded areas. Naturwissenschaften 2009, 96:279-288.
5.   Wyman CE: What is (and is not) vital to advancing cellulosic ethanol. Trends Biotechnol 2007, 25:153-157.
6.   Sjöström E, Alén R (Eds): Analytical methods in wood chemistry, pulping, and papermaking. Berlin, Heidelberg: Springer-Verlag; 1999.
7.   Rowell RM: Handbook of wood chemistry and wood composites. 2nd edition. Boca Raton FL: CRC Press; 2012.

8. Mabee WE, Gregg DJ, Arato C, Berlin A, Bura R, Gilkes N, Mirochnik O, Pan X, Pye EK, Saddler JN: Updates on softwood-to-ethanol process development. Appl Biochem Biotechnol 2006, 129–132:55-70.
9. Galbe M, Zacchi G: Pretreatment of lignocellulosic materials for efficient bioethanol production. Adv Biochem Eng Biotechnol 2007, 108:41-65.
10. Cullis IF, Mansfield SD: Optimized delignification of wood-derived lignocellulosics for improved enzymatic hydrolysis. Biotech Bioeng 2010, 106:884-893.
11. Shuai L, Yang Q, Zhu JY, Lu FC, Weimer PJ, Ralph J, Pan XJ: Comparative study of SPORL and dilute-acid pretreatments of spruce for cellulosic ethanol production. Bioresour Technol 2010, 101:3106-3114.
12. Arantes V, Saddler JN: Cellulose accessibility limits the effectiveness of minimum cellulase loading on the efficient hydrolysis of pretreated lignocellulosic substrates. Biotechnol Biofuels 2011, 4:3.
13. Viikari L, Alapuranen M, Puranen T, Vehmaanperä J, Siika-Aho M: Thermostable enzymes in lignocellulose hydrolysis. Adv Biochem Eng Biotechnol 2007, 108:121-145.
14. Arantes V, Saddler JN: Access to cellulose limits the efficiency of enzymatic hydrolysis: the role of amorphogenesis. Biotechnol Biofuels 2010, 3:4.
15. Van Zyl WH, Lynd LR, den Haan R, McBride JE: Consolidated bioprocessing for bioethanol production using Saccharomyces cerevisiae. Adv Biochem Eng Biotechnol 2007, 108:205-235.
16. Dunlop MJ: Engineering microbes for tolerance to nextgeneration biofuels. Biotechnol Biofuels 2011, 4:32.
17. Teixeira MC, Godinho CP, Cabrito TR, Mira NP, Sá-Correia I: Increased expression of the yeast multidrug resistance ABC transporter Pdr18 leads to increased ethanol tolerance and ethanol production in high gravity alcoholic fermentation. Microb Cell Fact 2012, 11:98.
18. Andrić P, Meyer AS, Jensen PA, Dam-Johansen K: Reactor design for minimizing product inhibition during enzymatic lignocellulose hydrolysis: I. Significance and mechanism of cellobiose and glucose inhibition on cellulolytic enzymes. Biotechnol Adv 2010, 28:308-324.
19. Podkaminer KK, Shao X, Hogsett DA, Lynd LR: Enzyme inactivation by ethanol and development of a kinetic model for thermophilic simultaneous saccharification and fermentation at 50°C with Thermoanaerobacterium saccharolyticum ALK2. Biotech Bioeng 2011, 108:1268-1278.
20. Bezerra RMF, Dias AA: Enzymatic kinetic of cellulose hydrolysis - Inhibition by ethanol and cellobiose. Appl Biochem Biotechnol 2005, 126:49-59.
21. Ximenes E, Kim Y, Mosier N, Dien B, Ladisch M: Inhibition of cellulases by phenols. Enzyme Microb Tech 2010, 46:170-176.
22. Martín C, Galbe M, Nilvebrant N-O, Jönsson LJ: Comparison of the fermentability of enzymatic hydrolysates of sugarcane bagasse pretreated by steam explosion using different impregnating agents. Appl Biochem Biotechnol 2002, 98–100:699-716.
23. Popoff T, Theander O: Formation of aromatic compounds from carbohydrates: Part III. Reaction of D-glucose and D-fructose in slightly acidic, aqueous solution. Acta Chem Scand 1976, 30:397-402.

24. Fenske JJ, Griffin DA, Penner MH: Comparison of aromatic monomers in lignocellulosic biomass prehydrolysates. J Ind Microbiol Biotechnol 1998, 20:364-368.

25. Chen SF, Mowery RA, Scarlata CJ, Chambliss CK: Compositional analysis of water-soluble materials in corn stover. J Agric Food Chem 2007, 55:5912-5918.

26. Du B, Sharma LN, Becker C, Chen S-F, Mowery RA, van Walsum GP, Chambliss CK: Effect of varying feedstock-pretreatment chemistry combinations on the formation and accumulation of potentially inhibitory degradation products in biomass hydrolysates. Biotech Bioeng 2010, 107:430-440.

27. Tran AV, Chambers RP: Red oak derived inhibitors in the ethanol fermentation of xylose by Pichia stipitis CBS 5776. Biotechnol Lett 1985, 7:841-845.

28. Clark TA, Mackie KL: Fermentation inhibitors in wood hydrolysates derived from the softwood Pinus radiata. J Chem Tech Biotechnol 1984, 34:101-110.

29. Tran AV, Chambers RP: Lignin and extractives derived inhibitors in the 2,3-butanediol fermentation of mannose-rich prehydrolysates. Appl Microbiol Biotechnol 1986, 23:191-197.

30. Ando S, Arai I, Kiyoto K, Hanai S: Identification of aromatic monomers in steam-exploded poplar and their influences on ethanol fermentation by Saccharomyces cerevisiae. J Ferment Technol 1986, 64:567-570.

31. Burtscher E, Bobleter O, Schwald W, Concin R, Binder H: Chromatographic analysis of biomass reaction products produced by hydrothermolysis of poplar wood. J Chromatogr 1987, 390:401-412.

32. Luo C, Brink DL, Blanch HW: Identification of potential fermentation inhibitors in conversion of hybrid poplar hydrolysate to ethanol. Biomass Bioenergy 2002, 22:125-138.

33. Larsson S, Reimann A, Nilvebrant N-O, Jönsson LJ: Comparison of different methods for the detoxification of lignocellulose hydrolysates of spruce. Appl Biochem Biotechnol 1999, 77:91-103.

34. Persson P, Larsson S, Jönsson LJ, Nilvebrant N-O, Sivik B, Munteanu F, Thörneby L, Gorton L: Supercritical fluid extraction of a lignocellulosic hydrolysate of spruce for detoxification and to facilitate analysis of inhibitors. Biotech Bioeng 2002, 79:694-700.

35. Persson P, Andersson J, Gorton L, Larsson S, Nilvebrant N-O, Jönsson LJ: Effect of different forms of alkali treatment on specific fermentation inhibitors and on the fermentability of lignocellulose hydrolysates for production of fuel ethanol. J Agric Food Chem 2002, 50:5318-5325.

36. Jönsson LJ, Palmqvist E, Nilvebrant N-O, Hahn-Hägerdal B: Detoxification of wood hydrolysates with laccase and peroxidase from the white-rot fungus Trametes versicolor. Appl Microbiol Biotechnol 1998, 49:691-697.

37. Klinke HB, Ahring BA, Schmidt AS, Thomsen AB: Characterization of degradation products from alkaline wet oxidation of wheat straw. Bioresour Technol 2002, 82:15-26.

38. Nilvebrant N-O, Reimann A, Larsson S, Jönsson LJ: Detoxification of lignocellulose hydrolysates with ion-exchange resins. Appl Biochem Biotechnol 2001, 91–93:35-49.

39. Alriksson B, Cavka A, Jönsson LJ: Improving the fermentability of enzymatic hydrolysates of lignocellulose through chemical in-situ detoxification with reducing agents.Bioresour Technol 2011, 102:1254-1263.
40. Lowry GH, Rosebrough NJ, Farr AL, Randall RJ: Protein measurement with the Folin phenol reagent. J Biol Chem 1951, 193:265-275.
41. Larsson S, Quintana-Sáinz A, Reimann A, Nilvebrant N-O, Jönsson LJ: Influence of lignocellulose-derived aromatic compounds on oxygen-limited growth and ethanolic fermentation by Saccharomyces cerevisiae. Appl Biochem Biotechnol 2000, 84:617-632.
42. Keweloh H, Weyrauch G, Rehm H-J: Phenol-induced membrane changes in free and immobilized Escherichia coli. Appl Microbiol Biotechnol 1990, 33:66-71.
43. Martín C, Galbe M, Wahlbom CF, Hahn-Hägerdal B, Jönsson LJ: Ethanol production from enzymatic hydrolysates of sugarcane bagasse using recombinant xylose-utilising Saccharomyces cerevisiae. Enzyme Microb Tech 2002, 31:274-282.
44. Chandel AK, Kapoor RK, Singh A, Kuhad RC: Detoxification of sugarcane bagasse hydrolysate improves ethanol production by Candida shehatae NCIM 3501. Bioresour Technol 2007, 98:1947-1950.
45. Jurado M, Prieto A, Martínez-Alcalá A, Martínez AT, Martínez MJ: Laccase detoxification of steam-exploded wheat straw for second generation bioethanol. Bioresour Technol 2009, 100:6378-6384.
46. Kim Y, Ximenes E, Mosier NS, Ladisch MR: Soluble inhibitors/deactivators of cellulase enzymes from lignocellulosic biomass. Enzyme Microb Tech 2011, 48:408-415.
47. Ulbricht RJ, Sharon J, Thomas J: A review of 5-hydroxymethylfurfural HMF in parental solutions. Fund Appl Toxicol 1984, 4:843-853.
48. Larsson S, Palmqvist E, Hahn-Hägerdal B, Tengborg C, Stenberg K, Zacchi G, Nilvebrant N-O: The generation of fermentation inhibitors during dilute acid hydrolysis of softwood. Enzyme Microb Tech 1999, 24:151-159.
49. Sárvári Horváth I, Sjöde A, Alriksson B, Jönsson LJ, Nilvebrant N-O: Critical conditions for improved fermentability during overliming of acid hydrolysates from spruce. Appl Biochem Biotechnol 2005, 121–124:1031-1044.
50. Pampulha ME, Loureiro-Dias MC: Combined effect of acetic acid, pH and ethanol on intracellular pH of fermenting yeast. Appl Microbiol Biotechnol 1989, 31:547-550.
51. Verduyn C, Postma E, Scheffers WA, Van Dijken JP: Physiology of Saccharomyces cerevisiae in anaerobic glucose-limited chemostat cultures. J Gen Microbiol 1990, 136:305-319.
52. Viegas CA, Sá-Correia I: Activation of plasma membrane ATPase of Saccharomyces cerevisiae by octanoic acid. J Gen Microbiol 1991, 137:645-651.
53. Verduyn C, Postma E, Scheffers WA, Van Dijken JP: Effect of benzoic acid on metabolic fluxes in yeast: a continuous-culture study on the regulation of respiration and alcoholic fermentation. Yeast 1992, 8:501-517.
54. Norman C, Howell KA, Millar AH, Whelan JM, Day DA: Salicylic acid is an uncoupler and inhibitor of mitochondrial electron transport. Plant Physiol 2004, 134:492-501.

55. Chung IS, Lee YY: Ethanol fermentation of crude acid hydrolyzate of cellulose using high-level yeast inocula. Biotech Bioeng 1985, 27:308-315.

56. Liu ZL, Slininger PJ, Dien BS, Berhow MA, Kurtzman CP, Gorsich SW: Adaptive response of yeast to furfural and 5-hydroxymethylfurfural and new chemical evidence for HMF conversion to 2,5-bis-hydroxymethylfuran. J Ind Microbiol Biotechnol 2004, 31:345-352.

57. Diaz De Villegas ME, Villa P, Guerra M, Rodríguez E, Redondo D, Martinez A: Conversion of furfural into furfuryl alcohol by Saccharomyces cerevisiae. Acta Biotechnol 1992, 12:351-354.

58. Sárvári Horváth I, Franzén CJ, Taherzadeh MJ, Niklasson C, Lidén G: Effects of furfural on the respiratory metabolism of Saccharomyces cerevisiae in glucose-limited chemostats. Appl Environ Microbiol 2003, 69:4076-4086.

59. Taherzadeh MJ, Gustafsson L, Niklasson C, Lidén G: Physiological effects of 5-hydroxymethylfurfural on Saccharomyces cerevisiae. Appl Microbiol Biotechnol 2000, 53:701-708.

60. Wahlbom CF, Hahn-Hägerdal B: Furfural, 5-hydroxymethyl furfural, and acetoin act as external electron acceptors during anaerobic fermentation of xylose in recombinant Saccharomyces cerevisiae. Biotech Bioeng 2002, 78:172-178.

61. Cavka A, Alriksson B, Ahnlund M, Jönsson LJ: Effect of sulfur oxyanions on lignocellulose-derived fermentation inhibitors. Biotech Bioeng 2011, 108:2592-2599.

62. Martinez A, Rodriguez ME, York SW, Preston JF, Ingram LO: Effects of Ca(OH)2 treatments ("overliming") on the composition and toxicity of bagasse hemicellulose hydrolysates. Biotech Bioeng 2000, 69:526-536.

63. Liu ZL: Molecular mechanisms of yeast tolerance and in situ detoxification of lignocellulose hydrolysates. Appl Microbiol Biotechnol 2011, 90:809-825.

64. Wadskog I, Adler L: Ion homeostasis in Saccharomyces cerevisiae under NaCl stress. In Yeast stress response. Edited by Hohmann S, Mager WH. Berlin: Springer-Verlag; 2003:201-240.

65. Helle S, Cameron D, Lam J, White B, Duff S: Effect of inhibitory compounds found in biomass hydrolysates on growth and xylose fermentation by a genetically engineered strain of S. cerevisiae. Enzyme Microb Tech 2003, 33:786-792.

66. Maiorella BL, Blanch HW, Wilke CR: Feed component inhibition in ethanolic fermentation by Saccharomyces cerevisiae. Biotech Bioeng 1984, 26:1155-1166.

67. Alexandre H, Charpentier C: Biochemical aspects of stuck and sluggish fermentation in grape must. J Ind Microbiol Biot 1998, 20:20-27.

68. Lin Y, Tanaka S: Ethanol fermentation from biomass resources: current state and prospects. Appl Microbiol Biotechnol 2006, 69:627-642.

69. Palmqvist E, Grage H, Meinander NQ, Hahn-Hägerdal B: Main and interaction effects of acetic acid, furfural, and p-hydroxybenzoic acid on growth and ethanol productivity of yeasts. Biotech Bioeng 1999, 63:46-55.

70. Zaldivar J, Ingram LO: Effect of organic acids on the growth and fermentation of ethanologenic Escherichia coli LY01. Biotech Bioeng 1999, 66:203-210.

71. Klinke HB, Olsson L, Thomsen AB, Ahring BK: Potential inhibitors from wet oxidation of wheat straw and their effect on ethanol production of Saccharomyces cerevisiae: wet oxidation and fermentation by yeast. Biotech Bioeng 2003, 81:738-747.

72. Olofsson K, Bertilsson M, Lidén G: A short review on SSF – an interesting process option for ethanol production from lignocellulosic feedstocks. Biotechnol Biofuels 2008, 1:7.
73. Pienkos PT, Zhang M: Role of pretreatment and conditioning processes on toxicity of lignocellulosic biomass hydrolysates. Cellulose 2009, 16:743-762.
74. Parawira W, Tekere M: Biotechnological strategies to overcome inhibitors in lignocellulose hydrolysates for ethanol production: review. Crit Rev Biotechnol 2011, 31:20-31.
75. Cantarella M, Cantarella L, Gallifuoco A, Spera A, Alfani F: Comparison of different detoxification methods for steam-exploded poplar wood as a substrate for the bioproduction of ethanol in SHF and SSF. Proc Biochem 2004, 39:1533-1542.
76. Alriksson B, Sjöde A, Sárvári Horváth I, Nilvebrant N-O, Jönsson LJ: Ammonium hydroxide detoxification of spruce acid hydrolysates. Appl Biochem Biotechnol 2005, 121–124:911-922.
77. Alriksson B, Sjöde A, Nilvebrant N-O, Jönsson LJ: Optimal conditions for alkaline detoxification of dilute-acid lignocellulose hydrolysates. Appl Biochem Biotechnol 2006, 129–132:599-611.
78. Ranatunga TD, Jervis J, Helm RF, McMillan JD, Wooley RJ: The effect of overliming on the toxicity of dilute acid pretreated lignocellulosics: The role of inorganics, uronic acids and ether-soluble organics. Enzyme Microb Tech 2000, 27:240-247.
79. Zhu J, Yong Q, Xu Y, Yu S: Detoxification of corn stover prehydrolyzate by trialkylamine extraction to improve the ethanol production with Pichia stipitis CBS 5776. Bioresour Technol 2011, 102:1663-1668.
80. Parajó JC, Dominguez H, Domínguez JM: Improved xylitol production with Debaryomyces hansenii Y-7426 from raw or detoxified wood hydrolysates. Enzyme Microb Tech 1997, 21:18-24.
81. Sárvári Horváth I, Sjöde A, Nilvebrant N-O, Zagorodni A, Jönsson LJ: Selection of anion exchangers for detoxification of dilute-acid hydrolysates from spruce. Appl Biochem Biotechnol 2004, 114:525-538.
82. Björklund L, Larsson S, Jönsson LJ, Reimann A, Nilvebrant N-O: Treatment with lignin residue - a novel method for detoxification of lignocellulose hydrolysates. Appl Biochem Biotechnol 2002, 98–100:563-575.
83. López MJ, Nichols NN, Dien BS, Moreno J, Bothast RJ: Isolation of microorganisms for biological detoxification of lignocellulosic hydrolysates. Appl Microbiol Biotechnol 2004, 64:125-131.
84. Nichols NN, Sharma LN, Mowery RA, Chambliss CK, van Walsum GP, Dien BS, Iten LB: Fungal metabolism of fermentation inhibitors present in corn stover dilute acid hydrolysate. Enzyme Microb Tech 2008, 42:624-630.
85. Palmqvist E, Hahn-Hägerdal B, Szengyel Z, Zacchi G, Rèczey K: Simultaneous detoxification and enzyme production of hemicellulose hydrolysates obtained after steam pretreatment. Enzyme Microb Tech 1997, 20:286-293.
86. Okuda N, Soneura M, Ninomiya K, Katakura Y, Shioya S: Biological detoxification of waste house wood hydrolysate using Ureibacillus thermosphaericus for bioethanol production. J Biosci Bioeng 2008, 106:128-133.

87. Wingren A, Galbe M, Zacchi G: Techno-economic evaluation of producing ethanol from softwood: comparison of SSF and SHF and identification of bottlenecks. Biotechnol Prog 2003, 19:1109-1117.

88. Wingren A, Galbe M, Roslander C, Rudolf A, Zacchi G: Effect of reduction in yeast and enzyme concentrations in a simultaneous-saccharification-and-fermentation-based bioethanol process. Appl Biotechnol Biochem 2005, 122:485-500.

89. Larsson S, Cassland P, Jönsson LJ: Development of a Saccharomyces cerevisiae strain with enhanced resistance to phenolic fermentation inhibitors in lignocellulose hydrolysates by heterologous expression of laccase. Appl Environ Microbiol 2001, 67:1163-1170.

90. Larsson S, Nilvebrant N-O, Jönsson LJ: Effect of overexpression of Saccharomyces cerevisiae Pad1p on the resistance to phenylacrylic acid and lignocellulose hydrolysates under aerobic and oxygen-limited conditions. Appl Microbiol Biotechnol 2001, 57:167-174.

91. Petersson A, Almeida JRM, Modig T, Karhumaa K, Hahn-Hägerdal B, Gorwa-Grauslund MF, Lidén G: A 5-hydroxymethyl furfural reducing enzyme encoded by the Saccharomyces cerevisiae ADH6 gene conveys HMF tolerance. Yeast 2006, 23:455-464.

92. Gorsich SW, Dien BS, Nichols NN, Slininger PJ, Liu ZL, Skory CD: Tolerance to furfural-induced stress is associated with pentose phosphate pathway genes ZWF1, GND1, RPE1, and TKL1 in Saccharomyces cerevisiae. Appl Microbiol Biotechnol 2006, 71:339-349.

93. Hasunuma T, Sanda T, Yamada R, Yoshimura K, Ishii J, Kondo A: Metabolic pathway engineering based on metabolomics confers acetic and formic acid tolerance to a recombinant xylose-fermenting strain of Saccharomyces cerevisiae. Microb Cell Fact 2011, 10:2.

94. Hasunuma T, Sung K, Sanda T, Yoshimura K, Matsuda F, Kondo A: Efficient fermentation of xylose to ethanol at high formic acid concentrations by metabolically engineered Saccharomyces cerevisiae. Appl Microbiol Biotechnol 2011, 90:997-1004.

95. Alriksson B, Sárvári Horváth I, Jönsson LJ: Overexpression of Saccharomyces cerevisiae transcription factor and multidrug resistance genes conveys enhanced resistance to lignocellulose-derived fermentation inhibitors. Proc Biochem 2010, 45:264-271.

96. Nevoigt E: Progress in metabolic engineering of Saccharomyces cerevisiae. Microbiol Mol Biol Rev 2008, 72:379-412.

97. Martinez A, Rodriguez ME, Wells ML, York SW, Preston JF, Ingram LO: Detoxification of dilute acid hydrolysates of lignocellulose with lime. Biotechnol Progr 2001, 17:287-293.

98. Van Zyl C, Prior BA, Du Preez JC: Production of ethanol from sugarcane bagasse hemicellulose hydrolyzate by Pichia stipitis. Appl Biochem Biotechnol 1988, 17:357-369.

99. Nilvebrant N-O, Persson P, Reimann A, de Sousa F, Gorton L, Jönsson LJ: Limits for alkaline detoxification of dilute-acid lignocellulose hydrolysates. Appl Biochem Biotechnol 2003, 105–108:615-628.

100. Millati R, Niklasson C, Taherzadeh MJ: Effect of pH, time and temperature of over-liming on detoxification of dilute-acid hydrolyzates for fermentation by Saccharomyces cerevisiae. Proc Biochem 2002, 38:515-522.
101. Mohagheghi A, Ruth M, Schnell DJ: Conditioning hemicellulose hydrolysates for fermentation: effects of overliming pH on sugar and ethanol yields. Proc Biochem 2006, 41:1806-1811.
102. Soudham VP, Alriksson B, Jönsson LJ: Reducing agents improve enzymatic hydrolysis of cellulosic substrates in the presence of pretreatment liquid. J Biotechnol 2011, 155:244-250.

PART III

ADVANCED MICROBIAL TECHNOLOGIES

CHAPTER 5

Microbial Production of Sabinene—A New Terpene-Based Precursor of Advanced Biofuel

HAIBO ZHANG, QIANG LIU, YUJIN CAO, XINJUN FENG, YANNING ZHENG, HUIBIN ZOU, HUI LIU, JIANMING YANG, AND MO XIAN

5.1 BACKGROUND

Progresses in metabolic engineering and synthetic biology boost the engineering of microbes to produce advanced biofuels [1-3]. Among the bio-based fuels, terpenes, which are derived from the head-to-tail condensation of dimethylallyl pyrophosphate (DMAPP) and isopentenyl pyrophosphate (IPP), and traditionally used in flavorings, fragrances [4], medicines and fine chemicals [5,6], have the potentials to serve as advanced biofuel precursors [7-9].

Terpenes are a large and diverse class of organic compounds, which are mainly produced by a variety of plants. They are generated from the common precursors, IPP and DMAPP, which can be produced from the methylerythritol 4-phosphate (MEP) pathway or the mevalonate (MVA)

Microbial Production of Sabinene—A New Terpene-Based Precursor of Advanced Biofuel. © 2014 Zhang et al.; licensee BioMed Central Ltd. Microbial Cell Factories 2014, 13:20 doi:10.1186/1475-2859-13-20. Creative Commons Attribution License (http://creativecommons.org/licenses/by/2.0).

pathway (Figure 1) [10]. Although many microorganisms harbor the MEP pathway or MVA pathway to supply the intermediates DMAPP and IPP, they are unable to produce the monoterpenes for the lack of monoterpenes synthases. With the rising demand for advanced fuels, terpene-based advance fuels attract more attentions. Many researchers explored microbial methods of monoterpene productions by introducing heterologous monoterpene synthase, including 3-carene, limonene, pinene and bisabolene. Reiling et al. engineered *E. coli* strain with overexpression native 1-deoxy-D-xylulose-5-phosphate synthase (DXS), farnesyl diphosphate synthase (IspA), IPP isomerase (IPIHp) from *Haematococcus pluvialis*, and 3-carene cyclase from *Picea abies*, which can accumulate a 3-carene titer of about 3 μg/L/OD$_{600}$ after 8 h production [11]. Carter et al. constructed a monoterpene biosynthesis pathway in *E. coli* with a titer of about 5 mg/L limonene production using the native MEP pathway [12]. Bisabolene, α-pinene et al. had been produced using MVA heterologous pathway in microorganisms [9,13,14].

Sabinene (CAS: 3387-41-5), a perfume additive, is being explored as the components for the next generation aircraft fuel [7,8]. Meanwhile, sabinene contributes to the spiciness of black pepper, is a principal component of carrot seed oil, and occurs at a low concentration in tea tree oil. Currently, sabinene is extracted from plants, which is inefficient and requires substantial expenditure of natural resources because of the low content of them [15]. Though sabinene was found in the culture of an endophytic *Phomopsis* sp. as a component of its volatile organic compounds, further work need to be done for microbial production method because of the low tilter in the mixture [16]. Consequently, green and sustainable microbial technologies, which could engineer microorganisms to convert renewable resources from biomass to biobased advanced biofuels, provided an alternative strategy [17-19].

In this paper, sabinene was significantly produced by assembling a biosynthetic pathway using the MEP or heterologous MVA pathway combining the GPP and sabinene synthase genes in an engineered *E. coli* strain. Subsequently, the culture medium and process conditions were optimized to enhance sabinene production. Finally, fed-batch fermentation of sabinene was evaluated using the optimized culture medium and process conditions.

5.2 RESULTS AND DISCUSSION

5.2.1 CHARACTERIZATION OF SABINENE BY GC-MS

E. coli cannot produce sabinene because of the absence of sabinene synthase, though it possesses a native MEP pathway which can supply the intermediates DMAPP and IPP (Figure 1). Consequently, sabinene synthase (SabS1) derived from *Salvia pomifera* was introduced into the *E.*

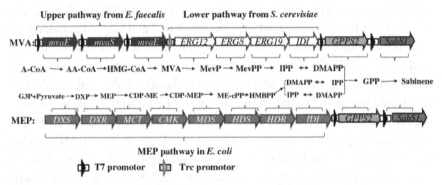

FIGURE 1: Sabinene biosynthesis pathway. Gene symbols and the enzymes they encode (all genes marked with black arrows were from *E. faecalis*, all genes marked with white arrows were isolated from *S. cerevisiae*, the gene marked with gray arrows and black characters were derived from *A. grandis* or *S. pomifera*, and the gene marked with gray arrows and white characters were native genes in *E. coli*). Enzymes in MVA pathway: MvaE, acetyl-CoA acetyltransferase/HMG-CoA reductase; MvaS, HMG-CoA synthase; ERG12, mevalonate kinase; ERG8, phosphomevalonate kinase; ERG19, mevalonate pyrophosphate decarboxylase; IDI, IPP isomerase; *A. grandis* geranyl diphosphate synthase (GPPS2) and *S. pomifera* sabinene synthase (SabS1) were optimized to the preferred codon usage of *E. coli*. Enzymes in MEP pathway: DXS, DXP synthase; DXR, DXP reductoisomerase; MCT, CDP-ME synthase; CMK, CDP-ME kinase; MDS, ME-cPP synthase; HDS, HMBPP synthase: HDR, HMBPP reductase; IDI, IPP isomerase. Intermediates in MVA pathway: A-CoA, acetyl-CoA; AA-CoA, acetoacetyl-CoA; HMG-CoA, hydroxymethylglutaryl-CoA; Mev-P, mevalonate 5-phosphate; Mev-PP, mevalonate pyrophosphate. IPP, isopentenyl pyrophosphate; DMAPP, dimethylallyl pyrophosphate; GPP, geranyl diphosphate. Intermediates in MEP pathway: G3P, glyceraldehyde 3-phosphate; DXP, 1-deoxy-D-xylulose 5-phosphate; MEP, 2-C-methyl-D-erythritol 4-phosphate; CDP-ME, 4-(cytidine-5′-diphospho)-2-C-methyl-D-erythritol; CDP-MEP, 2-phospho-4-(cytidine-5′-di-phospho)-2-C-methyl-D-erythritol; ME-cPP, 2-C-methyl-D-erythritol 2,4-cyclodiphosphate; HMBPP, 4-hydroxy-3-methylbut-2-enyl diphosphate; IPP, isopentenyl diphosphate; DMAPP, dimethylallyl diphosphate..

coli strain (HB1), to synthesize sabinene. However, after 36 h of incubation of the modified strain, only trace of the target product could be detected by GC-MS (data not shown), based on the relative retention time and total ion mass spectral comparison with the external standard. The main reason might lie in the insufficiency of GPP in the host, because the wild *E. coli* seldom produces terpene. Hence, the native gene IspA from *E. coli* W3100, which encodes farnesyl diphosphate synthase, was added to enhance the metabolic flux into GPP by catalyzing the conversion of DMAPP and IPP. The gene IspA combining with the sabinene synthase gene (SabS1) was ligated into pACYCDuet-1 to create the plasmid pHB3 (pACY-IspA-SabS1). The *E. coli* strain harboring pHB3 was inoculated in the initial fermentation medium and incubated at 37°C with shaking at 180 rpm in shake-flasks. IPTG was added to a final concentration of 0.5 mM when its OD600 reached 0.6-0.9, and culture was further maintained at 37°C for 24 h. The off-gas from the headspace of the sealed cultures was tested by GC-MS. The engineered *E. coli* BL21(DE3) strain harboring the native IspA gene and SabS1 from *S. pomifera* produced sabinene in detectable quantities (shown in Figure 2). Thus, using the MEP pathway and SabS1 from *S. pomifera*, the biosynthetic pathway for sabinene production was successfully constructed in *E. coli* BL21(DE3). The result also indicated that introduction of GPP synthase was beneficial to enhance the metabolic flux into GPP which would improve the sabinene products efficiently.

5.2.2 SCREENING OF GPP SYNTHASES

GPP synthase, is one of the rate-limiting enzyme in the sabinene synthesis of *E. coli* BL21(DE3) [20]. An effective method to optimize pathway efficiency may be to use genes of rate-limiting enzymes from different organisms [21]. In this study, GPPS enzymes from Abies grandis (GPPS2) and *E. coli* were evaluated to enhance the supply of GPP.

The GPPS2 gene from A. grandis or IspA gene from *E. coli* was cloned into the plasmid pACYCDuet-1 along with the sabinene synthase gene (SabS1) to create the plasmid pHB3 or pHB5, respectively, which were subsequently harbored by *E. coli* BL21(DE3) to screen the GPP synthase,

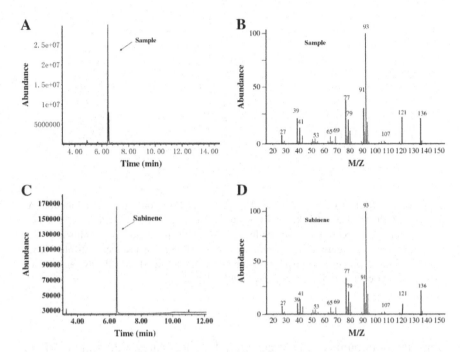

FIGURE 2: GC-MS analysis of sabinene from the headspace of the sealed cultures of strain HB2. Cultures were induced at 37°C, OD600=0.6-0.9, and final concentration of 0.25 mM IPTG. By comparing with the authoritative sabinene (A, B), the capacities of sabinene biosynthesis were verified. A, C, total ion chromatogram (TIC); C, D, mass spectrum. Based on the relative retention time and total ion mass spectral comparison with an external standard, sabinene production was identified.

because of the difficulty in detecting and quantifying GPP. The strains HB2 (harboring pHB3) and HB3 (harboring pHB5) were cultured in 600-ml shake-flasks with 100 ml of fermentation medium. When each culture reached an OD600 of 0.6-0.9, expression of GPP synthase and sabinene synthase was induced by 0.25 mM IPTG, and the culture was further incubated at 37°C for 24 h. A noticeable difference in sabinene production was observed between the two strains. The strain HB3 produced 2.07 mg/L sabinene, while the strain HB2 produced 0.96 mg/L (Figure 3A). This result demonstrates that the exogenous expression of GPPS contributed to the sabinene production, and the enzyme activity of GPPS2 from A. grandis was higher than that of IspA from *E. coli* W3100. IspA could give

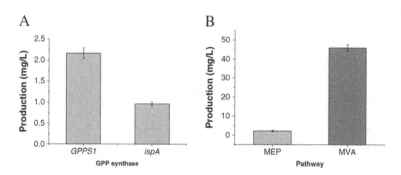

FIGURE 3: Effect of GPP synthase and metabolic pathway on sabinene production. A: Effect of GPP synthase. GPPS1, GPP synthase form A. grandis; IspA, GPP synthase from *E. coli*. B: Effect of metabolic pathway on sabinene production, the pathway details were described in Figure 1. The strain harboring GPPS2 can produce 2.2-fold higher concentration of sabinene than IspA, while the strain harboring MVA pathway can produce 20-fold higher concentration of sabinene than MEP pathway.

substantial amounts of the larger prenyl diphosphates, FPP and GGPP, in addition to GPP [22], that was why GPPS2 from *A. grandis* was more efficient than IspA in the synthesis of GPP. Hence, the GPPS2 was selected to enhance GPP production in the following experiments.

5.2.3 SCREENING OF SYNTHETIC PATHWAYS FOR SABINENE PRODUCTION

The hybrid exogenous MVA pathway is effective to synthesize DMAPP and IPP according to previous experimental data. The recombinant strain HB4 (*E. coli* harboring the MVA pathway, GPPS and sabinene synthase) and strain HB3 (*E. coli* harboring the native MEP pathway, GPPS synthase and sabinene synthase) were cultured to test the effect of the MVA pathway on the production of sabinene, in fermentation medium under shake-flask conditions. The sabinene titer of strain HB4 reached 44.74 mg/L after being induced by 0.25 mM IPTG for 24 h with glycerol as carbon source and beef powder as nitrogen source (Figure 3B). The titer was about 20-fold higher than that of the strain HB3 cultured at the same conditions.

These results indicated that the hybrid MVA pathway caused a huge increase in sabinene production, which was accordant with the production of other terpenes using a hybrid exogenous MVA pathway in engineered *E. coli* strains [23]. One reason for the inefficiency of MEP approach was the regulatory mechanisms present in the native host [24]. This limitation was also confirmed by experiments on isoprene production using the MEP or MVA pathway [23,25]. It is because the hybrid exogenous MVA pathway is effective to synthesize DMAPP and IPP, which are the precursors of GPP. Consequently, we hypothesized that the engineered strain with the hybrid exogenous MVA pathway could further enhance the production of sabinene. Therefore, the strain HB4 harboring the hybrid exogenous MVA pathway was chosen for further experiments.

5.2.4 OPTIMIZATION OF FERMENTATION MEDIUM AND CULTURE CONDITIONS

Fermentation medium and culture conditions play a vital role in the formation, concentration and yield of the end product [26], and they also provide data for fed-batch fermentation. Optimizing fermentation medium and culture conditions for strains can make the fed-batch fermentation easy to get higher quality and quantity products. In this study, the one-factor-at-a-time method, which is based on the classical method of changing one independent variable while fixing all others [27,28], is applied to optimize medium components as well as process conditions. The four most important factors, carbon source, organic nitrogen source, induction temperature, and inducer concentration were optimized to improve sabinene production, using the strain HB4.

5.2.5 EFFECT OF ORGANIC NITROGEN SOURCE ON SABINENE PRODUCTION

The source of the nitrogen in the medium, especially the organic, which can also provide trace nutrition for micro-being, plays an important role in improving the biosynthesis of desired products [29]. Four different organic nitrogen sources were assessed to investigate the effect of organic nitrogen source on sabinene production (Figure 4A). Beef powder permitted a little higher sabinene production than other organic nitrogen sources, among the organic nitrogen supplements tried. The highest concentration of sabinene was 23.20 mg/L, which was about 1.4 times as much as the lowest observed yeast extract.

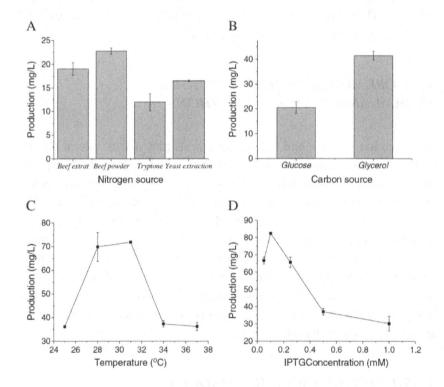

FIGURE 4: Effects of fermentation source and culture conditions on sabinene production by HB4. A: Effect of nitrogen sources on sabinene production; B: Effect of carbon sources on sabinene production; C: Effect of temperatures on sabinene production; D: Effect of the concentration of inducer on sabinene production. When OD600 reached 0.6-0.9, cultures were induced for 24 h using IPTG in shake-flasks. All the experiments were performed in triplicates. Optimized conditions: Nitrogen sources, beef power; Carbon source, glycerol; Temperature, 31°C; IPTC concentration, 0.1 mM.

5.2.6 EFFECT OF CARBON SOURCE ON SABINENE PRODUCTION

The source of carbon is the main feedstock in most fermentation media; therefore, finding efficient and cheap carbon source for sabinene production is important. In this study, mostly used carbon source glucose and glycerol were applied to investigate the effect of carbon source on sabinene production. As is shown in Figure 4B, the glycerol permitted a little higher sabinene production than glucose. The highest concentration of sabinene was 41.45 mg/L, which was about 2.03 times as much as that of glucose as carbon source.

5.2.7 EFFECT OF INDUCTION TEMPERATURE ON SABINENE PRODUCTION

In this study, the induction temperatures of 25°C, 28°C, 31°C, 34°C and 37°C were tried to increase sabinene production. As is shown in Figure 4C, the maximum sabinene production was observed at 31°C, at 71.50 mg/L with beef power as nitrogen source and glycerol as carbon source. It was about 2 times greater than that observed at 25°C (36.12 mg/L), and 37°C (36.19 mg/L). The enzyme expression, cell growth and product formation should be balanced in a successful control of cultivation temperature. Because low temperatures decrease the inclusion bodies in genetically engineered *E. coli*, the activities of recombinant enzymes can be enhanced by low induction temperatures (25°C or 30°C) [30,31]. Hence, the optimal induction temperature for sabinene production was around 31°C.

5.2.8 EFFECT OF INDUCER CONCENTRATION ON SABINENE PRODUCTION

To optimize the inducer concentration, various IPTG concentrations, ranging from 0.05 mM to 1 mM, were tested. The production of sabinene reached a maximum of 82.18 mg/L at the IPTG concentration of 0.1 mM, which was about 2.45 times greater than those observed at 1.0 mM (33.14

mg/L). The level of IPTG used can be varied to adjust the extent of the metabolic burden imposed on the cell [32], which can result in reduced growth rates, cell yields, protein expression, and plasmid stability [33,34].

Therefore, the most suitable medium was using beef powder and glycerol as the nitrogen and carbon source, respectively, and optimal culture temperature for sabinene production was 31°C at the concentration of 0.1 mM IPTG using the engineered strain HB4.

5.2.9 TOXICITY OF COMMERCIAL SABINENE TO E. COLI

Toxicity of sabinene to the overproducing organism plays an important role in the biosynthetic process. The commercial sabinene imparts toxicity to *E. coli* when added exogenously to the medium (Figure 5). The *E. coli* cell growth was comparable at 0 g/L and up to 5 g/L of exogenously added sabinene. Though the log phase of *E. coli* was prolonged with about 12 h by even 0.5 g/L sabinene compared with the control, it can grow in 5 g/L sabinene with inhibiting rates of 70% after inoculation and cultured for 36 hours (Figure 5). The results indicated that sabinene can be produced with the engineered *E. coli*, but the production tolerance cannot be neglected to get high titers.

The toxicity of products to the hosts is common in biosynthesis of biofuels and chemicals [3,35]. Expression of efflux pumps, heat shock proteins, membrane modifying proteins, and activation of general stress response genes all can improve tolerance of the hosts [2,3,36]. Furthermore, in situ product removal and membrane technology both can be used in the production of sabinene to get high titers [37].

5.2.10 FED-BATCH CULTURE OF THE ENGINEERED STRAINS

Fed-batch fermentation was carried out using the engineered *E. coli* BL21(DE3) strain simultaneously harboring plasmids pHB7 and pTrcLower, in the optimized medium and culture conditions, to further determine the ability of the engineered strain to produce sabinene at high yield. Glycerol was added continuously when the initial carbon source was exhausted

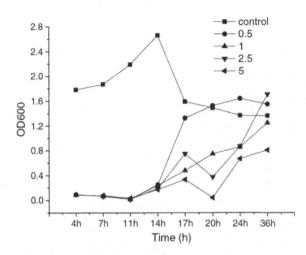

FIGURE 5: The growth of E. coli in LB medium with different concentration of commercial sabinene. The growth of the bacterial culture was determined by measuring the OD600 (the optical density at 600 nm) with a spectrophotometer (Cary 50 UV-Vis, Varian) at 4 h, 7 h, 11 h, 14 h, 17 h, 20 h, 24 h and 36 h. The concentrations of sabinene were added to the LB medium as follows: 0 g/L (■), 0.5 g/L (●), 1 g/L (▲), 2.5 g/L (▼), 5 g/L (◄).

which was indicated by the sharp rise of DO. As is shown in Figure 6, sabinene production increased rapidly from 4 h to 20 h after induction. After the cultures were induced for 24 h, sabinene reached a maximum concentration of 2.65 g/L with an average productivity of 0.018 g/h/g dry cells, and the conversion efficiency of glycerol to sabinene (gram to gram) reached 3.49%.

The maximum cell density of the engineered strain reached only about 14, four hours after being induced with IPTG, with a sabinene titer of no more than 0.5 g/L, which was rather low for the fed-batch fermentation of *E. coli* strains. The main reason for the low cell mass of *E. coli* strain may lie in the retardation of cell growth resulting from toxicity of the product, which was proved by the experiment of toxicity. Meanwhile, overexpression of many heterologous genes may be another reason. To resolve the above-mentioned problems, many possible improvements can be achieved to enhance sabinene production. One approach is to optimize the fermentation process by increasing cell density to elevate the yield of products

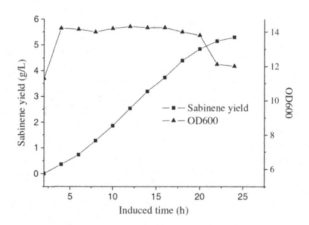

FIGURE 5: The time course of sabinene production by HB4 harboring pHB7 and pTrcLower in fed-batch fermentation. sabinene accumulation (■) and cell growth (▲). Induction was carried out at 12 h at the OD_{600} of 11. The maximum concentration of sabinene was 2.65 g/L with an average productivity of 0.018 g h^{-1} g^{-1} dry cells, and the conversion efficiency of glycerol to sabinene (gram to gram) was 3.49%.

[38,39], using in situ product removal, membrane technology or dissociation of growth or cell mass formation from product formation to reduce the toxicity of sabinene [37]. Another approach is engineering of the host including: employing a chromosome integration technique to decrease the cell growth burden on the host that results from overexpression of heterologous genes [9], expression of efflux pumps, heat shock proteins, membrane modifying proteins, and activation of general stress response genes to improve tolerance of the host to sabinene [2,3,36].

5.3 CONCLUSIONS

In this study, sabinene was significantly produced by assembling a biosynthetic pathway using the MEP or heterologous MVA pathway combining the GPP and sabinene synthase genes in an engineered *E. coli* strain. Subsequently, the culture medium and process conditions were optimized to enhance sabinene production. Finally, we also evaluated the fed-batch fermentation of sabinene using the optimized culture medium and process conditions, sabinene reached a maximum concentration of 2.65 g/L with an average productivity of 0.018 g/h/g dry cells, and the conversion efficiency of glycerol to sabinene (gram to gram) reached 3.49%. As far as we know, this is the first report of biosynthesis of sabinene using an engineered *E. coli* strain with the renewable carbon source as feedstock. Therefore, a green and sustainable production strategy has been provided for sabinene from renewable sources in *E. coli*.

5.4 METHODS AND MATERIALS

5.4.1 PLASMIDS, BACTERIAL STRAINS, AND GROWTH CONDITIONS

All plasmids and strains used in this study are listed in Table 1. *E. coli* BL21(DE3) (Invitrogen, Carlsbad, CA) was used as the host to overexpress proteins and produce sabinene. Cultures were grown aerobically at 37°C in Luria Broth (tryptone 10 g/L, NaCl 10 g/L, and yeast extract 5 g/L at pH 7.0-7.4). For initial production of sabinene experiments in shake-flasks, strains were grown in a medium (initial production medium) [40] consisting of the following: 20 g/L glucose, 9.8 g/L K_2HPO_4, 5 g/L beef extract, 0.3 g/L ferric ammonium citrate, 2.1 g/L citric acid monohydrate, 0.06 g/L $MgSO_4$ and 1 ml/L of trace element solution, which included (NH4)6Mo7O24·4H2O 0.37 g/L, $ZnSO_4$·7H2O 0.29 g/L, H3BO4 2.47 g/L, CuSO4·5H2O 0.25 g/L, and MnCl2·4H2O 1.58 g/L. Ampicillin (Amp, 100 μg/mL) and chloramphenicol (Cm, 34 μg/mL) was added if necessary.

Biosensor equipped with glucose oxidase membrane electrodes (Shandong Academy of Sciences, Jinan, China) was applied to determine the concentration of glucose.

TABLE 1: Plasmids and strains used in this study.

Name	Relevant characteristics	References
Plasmids		
pACYCDuet-1	P15A origin; Cm^R; PT7	Novagen
pTrcHis2B	ColE1 origin; Amp^R; P$_{trc}$	Invitrogen
pGH	pUC origin; Amp^R; P$_{T7}$	Generay
pTrcLower	ColE1 origin; Amp^R; P$_{trc}$:: ERG12-ERG8-ERG19-IDI1	[42]
pHB1	P15A origin; Cm^R; P$_{T7}$:: SabS1	This work
pHB2	P15A origin; Cm^R; P$_{T7}$::IspA	This work
pHB3	P15A origin; Cm^R; P$_{T7}$::IspA-SabS1	This work
pHB4	P15A origin; Cm^R; P$_{T7}$:: GPPS2	This work
pHB5	P15A origin; Cm^R; P$_{T7}$:: GPPS2-SabS1	This work
pHB6	P15A origin; Cm^R; P$_{T7}$:: mvaE-GPPS2-SabS1	This work
pHB7	P15A origin; Cm^R; P$_{T7}$::mvaE-mvaS-GPPS2-SabS1	This work
Strains		
E. coli BL21(DE3)	E. coli B dcm ompT hsdS(rB - mB -) gal	Takara
E. coli DH5α	deoR, recA1, endA1, hsdR17(rk-, mk+), phoA, supE44, λ-, thi-1, gyrA96, relA1	Invitrogen
Saccharomyces cerevisiae	Type strain	ATCC 204508
HB1	E. coli BL21(DE3) harboring pHB1	This work
HB2	E. coli BL21(DE3) harboring pHB3	This work
HB3	E. coli BL21(DE3) harboring pHB5	This work
HB4	E. coli BL21(DE3) harboring pHB7 and pTrcLower	This work

5.4.2 PLASMID CONSTRUCTION

The experiments were carried out according to standard protocols [41]. Polymerase chain reaction (PCR) was performed using Pfu DNA polymerase (TaKaRa, Dalian, China) according to the manufacturer's instructions.

5.4.3 CONSTRUCTION OF PLASMIDS
FOR GPP SYNTHASE SCREENING

E. coli BL21(DE3) genomic DNA was amplified as a template to obtain the IspA gene by PCR using the primers IspA-F and IspA-R (Table 2). The IspA gene fragment was digested using Bgl II and Nde I, and subsequently cloned into the corresponding sites of the vector pACYCDuet-1 to create pHB2 (Table 1). The SabS1 gene fragment (mentioned blow) was obtained by digestion of pGH/Pt30 with Bgl II and Xho I and was introduced into the corresponding sites of pHB2 to create pHB3.

The geranyl diphosphate synthase gene (GPPS2, GenBank No. AF513112) from *Abies grandis* and sabinene synthase gene (SabS1, GenBank No. ABH07678.1) from *Salvia pomifera* were analyzed by online software (http://www.genscript.com/cgi-bin/tools/rare_codon_analysis webcite) and optimized to the preferred codon usage of *E. coli* (http://www.jcat.de/ webcite). The codon-optimized GPPS2 gene and SabS1 gene were synthesized by Generay Company with plasmid pGH as the vector (pGH-GPPS2 and pGH-SabS1). The SabS1 gene fragment was obtained by digestion of pGH-SabS1 with Bgl II and Xho I and then cloned into the corresponding sites of pACYCDuet-1 to create pHB1.The GPPS2 gene fragment was obtained by digestion of pGH-GPPS2 with Nde I and Bgl II and then cloned into the corresponding sites of pACYCDuet-1 to create pHB4. The SabS1 gene fragment was obtained by digesting pGH-SabS1 with Bgl II and Xho I and was ligated into the corresponding sites of pHB4 to construct pHB5.

5.4.4 CONSTRUCTION OF PLASMIDS
FOR THE WHOLE PATHWAY OF SABINENE SYNTHESIS

As mentioned above, *E. coli* BL21(DE3) has its native MEP pathway to form IPP and DMAPP. Therefore, the MEP pathway for sabinene synthesis was constructed by harboring the plasmid pHB1 to introduce the exogenous sabinene synthase. Furthermore, to enhance the metabolic flux into GPP by catalyzing the conversion of DMAPP and IPP, the GPP synthase (IspA or GPPS2) was overexpressed or introduced.

TABLE 2: Primers used in this study.

Name	Sequence ($5' \rightarrow 3'$)
IspA-F	GGGAATTCCATATGATGGACTTTCCGCAGCAACTC
IspA-R	GGAAGATCTTTATTTATTACGCTGGATGATGT
mvaE-F	CATGCCATGGAGGAGGTAAAAAAACATGAAAACAG-TAGTTATTATTGATGC
mvaE-R	CGCGGATCCTTATTGTTTTCTTAAATCATTTAAAATAGC-GCGGA TCCTTATTGTTTTCTTAAATCATTTAAAATAG
mvaS-F	CCAGAGCTCAGGAGGTAAAAAAACAT-GACAATTGGGATTGATAAAATTA
mvaS-R	CAACTGCAGTTAGTTTCGATAAGAGCGAACG

E. coli BL21(DE3) harboring pHB7 and pTrcLower was constructed to form the MVA pathway for sabinene synthesis. The mvaE (Genbank: AF290092) was amplified with the primer mvaE-F and mvaE-R from genomic DNA of Enterococcus faecalis (ATCC 700802D-5) and then cloned into pHB5 and with restriction enzymes Nco I and Bam HI, creating pHB6. The mvaS (Genbank: AF290092) was amplified from genomic DNA of E. faecalis (ATCC 700802D-5) with the primer mvaS-F and mvaS-R and cloned into pHB6 and with restriction enzymes Sac I and Pst I, creating pHB7. The ERG12, ERG8, ERG19 and IDI1 genes from S. cerevisiae (ATCC 204508) were cloned into pTrcHis2B (Invitrogen, Carlsbad, CA) using a method of successive hybridization to yield pTrcLower [42].

5.4.5 CHARACTERIZATION OF SABINENE BY GC-MS

The *E. coli* strain was inoculated in 50 ml of fermentation medium containing 34 µg/mL Cm and then cultured at 37°C with shaking at 180 rpm. When the OD600 of the bacterial culture reached 0.6-0.9, the cells were induced by IPTG at a final concentration of 0.25 mM for 24 h. Then, the off-gas samples were taken from the headspace of the sealed cultures and analyzed by GC-MS.

Products characterization was carried out by capillary GC-MS using an Agilent 5975C system chromatograph. A HP-INNOWAX capillary column (30 m × 0.25 mm × 0.25 µm, Agilent, Palo Alto, CA, USA) was used, with helium as the carrier gas at a flow rate of 1 ml min-1. The following oven temperature program was carried out: 40°C for 1 min, increase of 4°C/min to 70°C, then programmed from 70°C to 250°C at 25°C/min, where it was held for 5 min. The injector temperature was maintained at 250°C; ion source temperature 230°C; EI 70 eV; mass range 35-300 m/z. suitable amount of samples were injected in split injection mode with a 20:1 split ratio. Peak identification was based on the relative retention time and total ion mass spectral comparison with the external standard.

5.4.6 QUANTIFICATION OF SABINENE BY GAS CHROMATOGRAPHY (GC)

The different strains were inoculated in 50 ml of fermentation medium containing 34 µg/mL Cm and/or 100 µg/mL Amp and then cultured under the conditions mentioned above. Finally, the off-gas samples were taken from the headspace of the sealed cultures and analyzed by GC.

The GC analysis was performed on an Agilent 7890A equipped with a flame ionization detector (FID). The separation of sabinene was performed using an HP-INNOWAX column (25 m × 250 µm × 0.2 µm). The linear velocity was 1 ml/min with N2 as carrier gas. The oven temperature was initially held at 50°C for 1 min, increased at 5°C/min to 100°C to 250°C, and finally held at 250°C for 5 min. The temperatures of injector and detector were held at 250°C and 260°C, respectively. The peak area was converted into sabinene concentration in comparison with a standard curve plotted with a set of known concentrations of sabinene which was bought from Sigma-Aldrich.

5.4.7 OPTIMIZATION OF FERMENTATION MEDIUM AND PROCESS

Optimization of fermentation medium was performed in shake-flask experiments in triplicate series of 600 ml sealed shake flasks containing 50 ml of fermentation medium incubated with the strain HB4. Amp (100 μg/mL) and Cm (34 μg/mL) were added when it was necessary. *E. coli* strains were cultured in the broth for initial production of sabinene and incubated in a gyratory shaker incubator at 37°C and 180 rpm. When the OD600 reached 0.6-0.9 [40], IPTG was added to a final concentration of 0.25 mM, and the culture was further incubated at 30°C for 24 h. Then, 1 ml of gas sample from the headspace of the sealed cultures was quantified as described previously [43]. Concentrations of synthesized sabinene were calculated by converting the GC peak area into milligrams of sabinene via a calibration curve.

5.4.8 EFFECT OF ORGANIC NITROGEN SOURCE

The shake-flask cultures were incubated in initial medium with different organic nitrogen sources (5 g/L): beef extract (solarbio), beef powder (MDBio, Inc), tryptone (Beijing AoBoXing Bio-Tech Co., Ltd) or yeast extract powder (Beijing AoBoXing Bio-Tech Co., Ltd)) at the above-mentioned culture conditions, and the sabinene products were detected.

5.4.9 EFFECT OF CARBON SOURCE

Carbon source is the main feedstock in fermentation. Therefore, the commonly used carbon sources (glucose and glycerol, 20 g/L) were screened in shake-flask with the nitrogen-optimized initial medium, at the above-mentioned culture conditions.

5.4.10 EFFECT OF INDUCTION TEMPERATURE

The *E. coli* strain was inoculated in 50 ml of optimized fermentation medium and cultured with shaking at 180 rpm. The shake-flask cultures were incubated at different induction temperatures (25°C, 28°C, 31°C, 34°C or 37°C), when the OD600 of the bacterial culture reached 0.6-0.9, for 24 h in a final concentration of 0.5 mM, and the sabinene products were quantified.

5.4.11 EFFECT OF IPTG CONCENTRATION

The shake-flask culture was incubated in different inducer (IPTG) concentrations (0.05 mM, 0.1 mM, 0.25 mM, 0.5 mM or 1 mM) at the optimized temperature for 24 h, and the sabinene products were measured.

5.4.12 TOXICITY OF COMMERCIAL SABINENE TO E. COLI

The sealed shake-flask culture was incubated in 50 ml of optimized fermentation medium and cultured with shaking at 180 rpm at a temperature of 31°C, with different concentration of commercial sabinene (0.5 g/L, 1 g/L, 2.5 g/L, and 5 g/L). Meanwhile, the growth of the bacterial culture was determined by measuring the OD600 (the optical density at 600 nm) with a spectrophotometer (Cary 50 UV-Vis, Varian) at 4 h, 7 h, 11 h, 14 h, 17 h, 20 h, 24 h and 36 h. The inhibition rate (IR) was calculated by the following equation:

$$IR = (1-OD_{600s}/OD_{600C}) \times 100\%$$

Where IR=Inhibition rate (100%); OD600s=OD600 of sample; OD600C=OD600 of control at the same time as the sample.

5.4.13 FED-BATCH FERMENTATION

The strain HB4 harboring pHB7 and pTrcLower was inoculated to 5 ml of LB medium (Amp 100 μg/mL, Cm 34 μg/mL, 37°C, 180 rpm), and then 100 ml fresh LB medium with corresponding antibiotics was inoculated with the 5 ml overnight cultures, which were used to inoculate a 5-L fermentor (BIOSTAT Bplus MO5L, Sartorius, Germany) containing 2 L of optimized fermentation medium (Amp 100 μg/mL, Cm 34 μg/mL). The temperature was maintained at 37°C firstly, and then 30°C after induced. The pH was maintained at 7.0 via automated addition of ammonia, and foam development was prohibited with 1% Antifoam 204. The stirring speed was first set at 400 rpm and then linked to the dissolved oxygen

(DO) concentration to maintain a 20% saturation of DO, the flow velocity of air was 1.5 L/min. The expression of heterogenous genes for sabinene production was initiated at an OD600 of 11 by adding IPTG at a final concentration of 0.15 mM, and IPTG was supplemented every 8 h. During the course of fermentation, the 40% glycerol was fed at a rate 4 g/L/h. Then, sabinene accumulation was measured every 60 min by GC as described above. Meanwhile, the growth of the bacterial culture was determined by measuring the OD600 with the spectrophotometer, and the dry cell weight was calculated according to the coefficient (one OD600 unit corresponded to 0.43 g/L of dry cell weight).

The specific productivity was calculated by the following equation [44].

$$Q_s = \frac{s_1 - s_0}{t_1 - t_0} \times \frac{2}{x_1 + x_0}$$

Where Qs=specific production rate (g/h/g dry cells); s=sabinene concentration (g/L); t=cultivation time (h), and x=biomass (g/L).

Conversion efficiency (gram to gram) of glycerol to sabinene was calculated by the following equation:

$$Y = G_s/G_g \times 100\%$$

Where Y=conversion efficiency (gram to gram, 100%); Gs=weight of sabinene (g); Gg=weight of glycerol (g).

5.4 ABBREVIATIONS

Amp: Ampicillin; Cm: Chloramphenicol; DMAPP: Dimethylallyl pyrophosphate; IPP: Isopentenyl pyrophosphate; GPP: Geranyl diphosphate; GPPS: Geranyl diphosphate synthase; MVA: Mevalonate; MEP: Methylerythritol 4-phosphate; IPTG: Isopropyl β-D-thiogalactoside; PCR: Polymerase chain reaction; GC: Gas chromatography; GC-MS: Gas chromatography-mass spectrography; DO: Dissolved oxygen.

REFERENCES

1. Peralta-Yahya PP, Keasling JD: Advanced biofuel production in microbes. Biotech J 2010, 5:147-162.
2. Piper PW: The heat shock and ethanol stress responses of yeast exhibit extensive similarity and functional overlap. FEMS Microbiol Lett 1995, 134:121-127.
3. Jiang X, Zhang H, Yang J, Liu M, Feng H, Liu X, Cao Y, Feng D, Xian M: Induction of gene expression in bacteria at optimal growth temperatures. Appl Microbiol Biotechnol 2013, 97(12):5423-5431.
4. Behr A, Johnen L: Myrcene as a natural base chemical in sustainable chemistry: a critical review. ChemSusChem 2009, 2:1072-1095.
5. Brown HC, Ramachandran PV: Asymmetric reduction with chiral organoboranes based on. alpha.-pinene. Accounts Chem Res 1992, 25:16-24.
6. Kirby J, Keasling JD: Biosynthesis of plant isoprenoids: perspectives for microbial engineering. Annu Rev Plant Biol 2009, 60:335-355.
7. Ryder JA: Patent US 7,935,156 B2 - Jet fuel compositions and methods of making and using same. 2011.
8. Rude MA, Schirmer A: New microbial fuels: a biotech perspective. Curr Opin Microbiol 2009, 12:274-281.
9. Chen H-T, Lin M-S, Hou S-Y: Multiple-copy-gene integration on chromosome of Escherichia coli for beta-galactosidase production. Korean J Chem Eng 2008, 25:1082-1087.
10. Steinbüchel A: Production of rubber-like polymers by microorganisms. Curr Opini Microbiol 2003, 6:261-270.
11. Reiling KK, Yoshikuni Y, Martin VJJ, Newman J, Bohlmann J, Keasling JD: Mono and diterpene production in Escherichia coli. Biotechnol Bioeng 2004, 87:200-212.
12. Carter OA, Peters RJ, Croteau R: Monoterpene biosynthesis pathway construction in Escherichia coli. Phytochemistry 2003, 64:425-433.
13. Peralta-Yahy PP, Ouellet M, Chan R, Mukhopadhyay A, Keasling JD, Lee TS: Identification and microbial production of a terpene-based advanced biofuel. Nat commun 2011, 2:483.
14. Bokinsky G, Peralta-Yahya PP, George A, Holmes BM, Steen EJ, Dietrich J, Lee TS, Tullman-Ercek D, Voigt CA, Simmons BA, Keasling JD: Synthesis of three advanced biofuels from ionic liquid-pretreated switchgrass using engineered Escherichia coli. Proc Natl Acad Sci U S A 2011, 108:19949-19954.
15. Chang MCY, Keasling JD: Production of isoprenoid pharmaceuticals by engineered microbes. Nat Chem Biol 2006, 2:674-681.
16. Singh SK, Strobel GA, Knighton B, Geary B, Sears J, Ezra D: An endophytic Phomopsis sp. possessing bioactivity and fuel potential with its volatile organic compounds. Microb Ecol 2011, 61:729-739.
17. Schmidt-Dannert C, Umeno D, Arnold FH: Molecular breeding of carotenoid biosynthetic pathways. Nat Biotechnol 2000, 18:750-753.
18. Clomburg JM, Gonzalez R: Biofuel production in Escherichia coli: the role of metabolic engineering and synthetic biology. Appl Microbiol Biotechnol 2010, 86:419-434.

19. Leonard E, Lim K-H, Saw P-N, Koffas MAG: Engineering central metabolic pathways for high-level flavonoid production in Escherichia coli. Appl Environ Microbiol 2007, 73:3877-3886.
20. Burke C, Croteau R: Geranyl diphosphate synthase from Abies grandis: cDNA isolation, functional expression, and characterization. Arch Biochem Biophys 2002, 405:130-136.
21. Yan YC, Liao J: Engineering metabolic systems for production of advanced fuels. J Ind Microbiol Biotechnol 2009, 36:471-479.
22. Fujisaki S, Nishino T, Katsuki H: Isoprenoid synthesis in Escherichia coli. Separation and partial purification of four enzymes involved in the synthesis. J Biochem 1986, 99:1327-1337.
23. Yang J, Xian M, Su S, Zhao G, Nie Q, Jiang X, Yanning Z, Liu W: Enhancing production of bio-isoprene using hybrid MVA pathway and isoprene synthase in E. coli. PloS One 2012, 7:e33509.
24. Martin VJJ, Pitera DJ, Withers ST, Newman JD, Keasling JD: Engineering a mevalonate pathway in Escherichia coli for production of terpenoids. Nat Biotechnol 2003, 21:796-802.
25. Zhao Y, Yang J, Qin B, Li Y, Sun Y, Su S, Xian M: Biosynthesis of isoprene in Escherichia coli via methylerythritol phosphate (MEP) pathway. Appl Microbiol Biotechnol 2011, 90:1915-19922.
26. Schmidt FR: Optimization and scale up of industrial fermentation processes. Appl Microbiol Biotechnol 2005, 68:425-435.
27. Ahamad M, Panda B, Javed S, Ali M: Production of mevastatin by solid-state fermentation using wheat bran as substrate. Res J Microbiol 2006, 1:443-447.
28. Alexeeva YV, Ivanova EP, Bakunina IY, Zvagintseva TN, Mikhailov VV: Optimization of glycosidases production by Pseudoalteromonas issachenkonii KMM 3549T. Lett Appl Microbiol 2002, 35:343-346.
29. Torija MJ, Beltran G, Novo M, Poblet M, Rozès N, Guillamón JM, Mas A: Effect of the nitrogen source on the fatty acid composition of Saccharomyces cerevisiae. Food Microbiol 2003, 20:255-258.
30. Hunke S, Betton J-M: Temperature effect on inclusion body formation and stress response in the periplasm of Escherichia coli. Mol Microbiol 2003, 50:1579-1589.
31. Groota NS, Ventura S: Effect of temperature on protein quality in bacterial inclusion bodies. FEBS Lett 2006, 580:6471-6476.
32. Donovan R, Robinson C, Glick B: Review: optimizing inducer and culture conditions for expression of foreign proteins under the control of thelac promoter. J Ind Microbiol Biotechnol 1996, 16:145-154.
33. Bentley WE, Mirjalili N, Andersen DC, Davis RH, Kompala DS: Plasmid-encoded protein: the principal factor in the "metabolic burden" associated with recombinant bacteria. Biotechnol Bioeng 1990, 35:668-681.
34. Glick BR: Metabolic load and heterologous gene expression. Biotechnol Adv 1995, 13:247-261.
35. Dunlop MJ: Engineering microbes for tolerance to next-generation biofuels. Biotechnol Biofuels 2011, 4:32.

36. Holtwick R, Meinhardt F, Keweloh H: cis-trans isomerization of unsaturated fatty acids: cloning and sequencing of the cti gene from Pseudomonas putida P8. Appl Environ Microbiol 1997, 63:4292-4297.
37. Ataei SA, Vasheghani-Farahani E: In situ separation of lactic acid from fermentation broth using ion exchange resins. J Ind Microbiol Biotechnol 2008, 35:1229-1233.
38. Wang D, Li Q, Song Z, Zhou W, Su Z, Xing J: High cell density fermentation via a metabolically engineered Escherichia coli for the enhanced production of succinic acid. J Chem Technol Biotechnol 2011, 86:512-518.
39. Chen N, Huang J, Feng Z, Yu L, Xu Q, Wen T: Optimization of fermentation conditions for the biosynthesis of L-Threonine by Escherichia coli. Appl Biochem Biotechnol 2009, 158:595-604.
40. Yang J, Zhao G, Sun Y, Zheng Y, Jiang X, Liu W, Xian M: Bio-isoprene production using exogenous MVA pathway and isoprene synthase in Escherichia coli. Bioresour Technol 2012, 104:642-647.
41. Sambrook J, Russell DW: Molecular cloning: a laboratory manual. 3rd edition. Cold Spring Harbor: Cold Spring Harbor Laboratory Press; 2001.
42. Jiang X, Yang J, Zhang H, Zou H, Wang C, Xian M: In vtro assembly of multiple DNA fragments using successive hybridization. PLoS ONE 2012, 7:e30267.
43. Kolb B: Headspace sampling with capillary columns. J Chromatogr A 1999, 842:163-205.
44. Tashiro Y, Takeda K, Kobayashi G, Sonomoto K, Ishizaki A, Yoshino S: High butanol production by Clostridium saccharoperbutylacetonicum N1-4 in fed-batch culture with pH-stat continuous butyric acid and glucose feeding method. J Biosci Bioeng 2004, 98:263-268.

CHAPTER 6

From Biodiesel and Bioethanol to Liquid Hydrocarbon Fuels: New Hydrotreating and Advanced Microbial Technologies

JUAN CARLOS SERRANO RUIZ, ENRIQUE V. RAMOS-FERNÁNDEZ AND ANTONIO SEPÚLVEDA-ESCRIBANO

6.1 INTRODUCTION

Petroleum, natural gas and coal, the so-called fossil fuels, supply most of the energy consumed worldwide and their exploitation has allowed our society to reach unprecedented levels of development during the past century. However, the large-scale consumption of these natural resources is associated with a number of important issues. First, combustion of fossil fuels for energy production releases large amounts of CO_2 (a greenhouse gas, GHG) into the atmosphere. This anthropogenic CO_2 cannot be fixed by plants at the current rates at which it is evolved thereby leading to accumulation and global warming. [1] In this sense, recent studies estimate that the burning of fossil fuels is responsible for 70% of the global warming problem. [2] Second, the uneven geographical distribution of fossil

From biodiesel and bioethanol to liquid hydrocarbon fuels: new hydrotreating and advanced microbial technologies. Energy & Environmental Science, 2012, 5, 5638-5652. DOI: 10.1039/C1EE02418C. Reproduced with permission from the Centre National de la Recherche Scientifique (CNRS) and The Royal Society of Chemistry.

fuel reserves, which in some cases are located in socio-politically unstable regions, is the origin of multiple political and economic issues worldwide, and obligates transportation of the fossil fuel over long distances to ensure supply to non-producer countries. Third, fossil fuel reserves are becoming less accessible with time, and their current consumption rate—boosted by the growing needs of industrialized countries and the rapid development of emerging economies—is, by far, higher than the natural regeneration cycle which inevitably leads to depletion within a few decades. [3]

In order to address these important concerns, a series of initiatives spurred by governments are trying to develop alternatives to fossil fuels which can progressively displace these nonrenewable resources in our current energy system. In this sense, a range of well-distributed carbon-free renewable sources such as solar, wind, hydroelectric and geothermal activity can substitute natural gas and coal in the production of heat and electricity, while biomass, the only sustainable source of organic carbon in earth, has been pointed out as the perfect equivalent to petroleum for the production of fuels, chemicals and carbonbased materials. [4]

Petroleum is the world's primary source of energy and chemicals and, among fossil fuels, is the resource with the shortest expected lifetime. More than 80 millions of barrels are consumed worldwide on a daily basis, and projections indicate that demand for crude oil will increase by 30% within the next 20 years. [5] A large fraction of the extracted crude (70–80%) ends up as fuels after processing in refineries to cover the elevated demand of the transportation sector, which is the largest and fastest growing energy sector and it is responsible for almost one third of the total energy consumed worldwide. [1] An eventual replacement of oil by biomass will thus necessarily involve the development of new technologies for the large-scale production of fuels from this resource, the so-called biofuels.

Important environmental (biofuels are considered as carbon neutral since the CO_2 produced during fuel combustion is consumed by subsequent biomass regrowth) and economic (reduction of the dependence on the strong fluctuations in the price of the oil, creation of new well-paid jobs in different sectors, revitalization of traditionally deprived rural areas) benefits are derived from the establishment of a solid biofuel industry. [6] However, the utilization of limited edible biomass feedstocks (e.g. sugars, starches and vegetable oils) for biofuels production leads to competition with food for land use, and research is now more focused on the utilization of more abundant and non-edible biomass (e.g. lignocellulose, waste oils, algae) which would allow sustainable production of biofuels without affecting food supplies or forcing changes in land use.

At the present time, the liquid biofuels most widely used are crop-based bioethanol and biodiesel, which have been successfully implemented in the transportation sector as alternatives to petrol-based gasoline and diesel, respectively. Biodiesel and bioethanol are known as conventional biofuels, that is, biofuels produced by simple and well-established technologies that are already generating fuels on a commercial scale. Conventional biofuels production has increased exponentially in the last few years, and the key fact for this rapid expansion is the partial compatibility of these biofuels with transportation infrastructure of diesel and gasoline which has permitted an easy penetration in the current fuel market.

Bioethanol is, by far, the predominant biomass-derived fuel at the present time, and only two countries U.S. (corn-derived) and Brazil (sugar cane-derived) monopolize more than 90% of the world's production. Ethanol is used as a high-octane additive of gasoline thereby improving combustion of the mixture. Furthermore, apart from the CO_2 emission savings, bioethanol usage allows reduction of pollutants such as CO, NO_x and SO_x. Fig. 1 shows a summarized scheme of the technology used for the production of bioethanol. The process starts with the extraction and deconstruction of the carbohydrate polymers forming part of the structure of biomass, which is typically carried out in water medium. This step is readily accomplished with edible biomass feedstocks such as sugar cane or corn (although in this last case an extra saccharification step is required), while the recalcitrant nature of lignocellulose obligates to perform costly pretreatment and hydrolysis steps which add complexity and increase production costs for biofuels derived from this resource.8 The released aqueous sugar monomers are subsequently fermented to the desired ethanol product (along with CO_2) using a variety of microorganisms (e.g., yeast, bacteria and mold), in a well-known process similar to that used in beer and wine-making. The stringent conditions required for microorganisms to survive obligate fermentation to be carried out at mild temperatures (e.g. 30–50°C) and low ethanol concentrations (lower than 15 wt%). Consequently, dilute aqueous solutions of ethanol are obtained

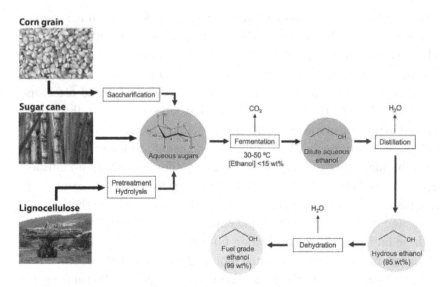

FIGURE 1: Summarized scheme of the fermentation technology utilized for the production of bioethanol from biomass feedstocks.

after fermentation, and an expensive energy-consuming distillation followed by additional dehydration steps (typically carried out over molecular sieves) are necessary to completely remove water from the mixture and reach the fuel-grade concentration level (e.g. ≥99 wt%). This deep water removal step, which typically accounts for 35–40% of the total energy required for bioethanol production, [9] is an inherent drawback associated with classical fermentation technologies. As described in a section below, new fermentation approaches can overcome this important limitation by converting sugars into hydrophobic hydrocarbons instead of water-soluble ethanol. Thus, liquid hydrocarbons spontaneously separate from the aqueous broth avoiding poisoning of bacteria by accumulated products and facilitating enormously separation/collection of the biofuel product.

Even though bioethanol is the predominant biofuel today, it has important compatibility, energy-density and water-absorption issues that limit its further implementation in the current fuel infrastructure. [10,11] These limitations are ultimately derived from the special physical-chemical properties of this compound. Thus, the corrosive nature of ethanol avoids its use as a pure fuel in current spark ignition engines which only tolerate low concentration blends (5–15%, v/v) with conventional gasoline (i.e., E5–E15), and additional engine upgrades, the so-called flexi fuel vehicles (FFVs), are required for ethanol enriched mixtures (E85). This constraint in ethanol blending, denoted as blend wall, is currently causing important issues in the U.S. to absorb the growing bioethanol production, and important figures have remarked that the U.S. will not be able to absorb the amount of renewable ethanol mandated by the Renewable Fuel Standard directive [12] ($1.4x10^{11}$ L by 2022) because of the lack of FFVs or E85 stations. [13,14]

Energy density, an important characteristic of any kind of fuel, decreases with the oxygen content in the molecule. Thus, ethanol (23.4 MJ L^{-1}) contains less energy per volume than conventional gasoline (34.4 MJ L^{-1}), which penalizes the fuel economy of cars running on E mixtures. In this sense, it is estimated that cars using E85 operate with 30% lower fuel mileage than those using regular gasoline. [15] This fact, if it is not adequately compensated with an equivalent reduction in the selling price of E blends (as happens in countries like Brazil), represents an important disadvantage that has discouraged drivers from purchasing E85 cars or fuel so far.

Fuels that do not absorb water are highly desirable. However, ethanol is highly hygroscopic and completely miscible in water. As a result, its addition to gasoline markedly increases the water solubility of the mixture and E blends are prone to suffering water contamination. When the water saturation level for a determined blend is reached, phase-separation episodes between water–ethanol and gasoline can occur causing important damage to the engine. [16]

Biodiesel is the second most abundant renewable liquid fuel (and the most common in Europe) with an annual production that reached $1.6x10^{10}$ L in 2009 and projections to triplicate this amount by 2020. [17] Biodiesel is nontoxic and sulfur free, and its use allows reduction of important pollutants of diesel engines such as particulate matter and hydrocarbon (although it increases NO_x emissions). [18] The conventional methodology for the production of biodiesel (Fig. 2) uses edible biomass feedstocks rich in oils such as palm, sunflower, canola, rapeseed and soybean. [19] Feedstocks such as low-quality waste oils and nonedible plants such as *Jatropha curcas* or *Camelina* are preferred since they are cheaper and do not contribute to increase competition with food supply. Recently, algae, with excellent oil contents (up to 75 wt%) and minimum land area utilization, have also been proposed as plausible oil sources for biodiesel production. [20]

The first step for any biodiesel technology involves oil extraction from the biomass source. This step is relatively wellestablished for edible feed-stocks and more troublesome for waste oils (the presence of water and free fatty acid impurities) and algae (lack of efficient methodologies for oil extraction). Vegetable oils, which are rich in triglycerides (TGs), are subsequently treated with methanol under mild temperatures (50–80°C) and in the presence of a basic homogeneous catalyst (Fig. 2). The process is known as transesterification and it allows conversion of TGs in a mixture of fatty acid methyl esters (FAME, the components of biodiesel) and glycerol (1,2,3-propanetriol). A large part of this co-produced glycerol is separated from FAME by simple decantation, although further washing/drying steps are required to remove traces of glycerol in order to comply with strict regulations for fuel grade biodiesel. This extra purification process increases production costs and generates great amounts of salts, soaps and waste water. Furthermore, the management of the large amounts of residual crude glycerol produced (100 kg per ton of biofuel) represents an important challenge for the biodiesel industry.

Biodiesel shares some of the compatibility and energy-density drawbacks of ethanol. It is slightly corrosive and, consequently, it can damage rubber and other components in the engine or fuel lines. [21] To avoid this problem, biodiesel usage in current vehicles is limited to low-concentration mixtures with conventional diesel fuel, the so-called B blends. Biodiesel, although in a lower extent than ethanol with gasoline, contains less energy per volume than regular diesel (34.5 vs. 40.3 MJ L^{-1}) and a slight gas mileage penalty is applied for vehicles running on B blends (2% for B20). [15] Another important disadvantage of biodiesel is its higher cloud point compared to regular diesel which increases the risk of plugging filters or small orifices at cold temperatures. [22] The operating conditions required for the new generation of diesel engines to increase efficiency (e.g. higher injection pressures and nozzles with a lower diameter) aggravate this issue. [23]

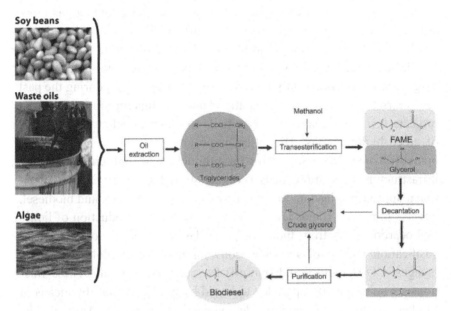

FIGURE 2: Summarized scheme of the transesterification process utilized for the production of biodiesel.

The current transportation infrastructure, including engines, fueling stations, distribution networks, and petrochemical technologies is entirely developed for petroleum-derived liquid hydrocarbons. These compounds are worth such a huge effort since they offer clean combustions, high energy densities, and superior stabilities, characteristics highly appreciated for a transportation liquid fuel. Even though bioethanol and biodiesel have found some room in the hydrocarbon-based infrastructure, there are important limitations (outlined above) that avoid a further penetration of these biofuels in the current fuel market. Remarkably, these limitations are ultimately derived from the different chemical compositions of these molecules compared to hydrocarbon fuels. This simple analysis has convinced many researchers around the world to explore new routes for the conversion of biomass into liquid hydrocarbon fuels chemically identical to those being used today in the transportation fleet. [10,11,24,25] Unlike biodiesel and bioethanol, those green hydrocarbons: (i) would not need to modify the existing infrastructure for their implementation in the transportation sector and could make use of the existing petroleum refinery facilities, (ii) would offer equivalent energy-density characteristics avoiding gas mileage penalties, and (iii) would overcome intrinsic drawbacks of bioethanol and biodiesel such as water absorption and high cloud point, respectively.

These strong incentives of green hydrocarbons have not been ignored by governments. As recently remarked by Regalbuto, [22] during the past few years there has been a dramatic change in funding directions from projects involving conventional biofuels to those aimed at the synthesis of green hydrocarbons. This effort is progressively favoring the development of new technologies for the production of high energy-density, infrastructure-compatible fuels (i.e. advanced biofuels) versus conventional approaches leading to oxygenates such as bioethanol and biodiesel. Several important routes are available today for the production of liquid hydrocarbon fuels from biomass. [11] Classical thermal routes such as gasification and pyrolysis allow conversion of lignocellulose into gas (syngas, CO/H_2) and liquid (bio-oil) fractions, respectively. These fractions are subsequently upgraded to liquid hydrocarbon fuels by means of Fischer–Tropsch and catalytic deoxygenation processes. Alternatively, soluble sugars, produced after pretreatment/hydrolysis of lignocellulose or by simple solubilization of starchy materials, can be transformed into gasoline, diesel and jet fuel by means of aqueous-phase catalytic routes involving deoxygenation and C–C coupling reactions. [7] Recently, two new promising approaches have been developed with the aim of producing green hydrocarbons from classical biomass feedstocks such as TGs and sugars: hydrotreating and advanced microbial synthesis. These new technologies, which can be seen as the advanced versions of classical transesterification and fermentation, will be the focus of the present paper. While gasification, pyrolysis and aqueous-phase processing are routes with a great potential, a detailed description of these technologies are beyond the scope of this perspective.

6.2 HYDROTREATING OF PLANT LIPIDS

Vegetable oils and related feedstocks can be transformed into liquid alkanes suitable for diesel and jet fuel applications in a process commonly denominated as hydrotreating. The process typically involves utilization of hydrogen at high pressures and moderate temperatures in the presence of supported metal catalysts.

The main reaction pathways involved in the hydrotreating of TGs to green hydrocarbons are shown in Fig. 3. The olefinic bonds typically present in TGs are readily hydrogenated at hydrotreating conditions. Once formed, the saturated TGs undergo breaking of the C–O bonds that maintain the tri-chain structure together by reaction with hydrogen (e.g. hydrogenolysis) thereby releasing the fatty acids along with a molecule of propane. These free fatty acids can then suffer oxygen removal by means of two different pathways: (i) repetitive cycles of hydrogenation and dehydration (e.g. hydrodeoxygenation, HDO) that progressively achieve aldehydes/ enols, [26] alcohols and, finally, n-alkanes with the same number of carbon atoms than the original fatty acids and; (ii) decarbonylation/ decarboxylation (HDC) that produces a linear alkane with n − 1 carbon atoms and the release of CO_x species. The theoretical maximum yield of hydrocarbons for HDO and HDC routes is 86 and 81 wt%, respectively. We note that, although Fig. 3 shows that hydrogen is involved in HDC, this process does not necessarily require the presence of this gas. Finally, the straight-chain alkanes can undergo further reactions such as isomerization to produce iso-alkanes and cracking to lower alkanes. While the former process is desirable to produce branched hydrocarbons with lower pour and cloud points (valuable for jet fuel and/or winter-diesel), cracking reactions should be minimized in the hydrotreating reactor since they typically cause catalyst deactivation and lead to lower-value small alkanes that decrease the cetane number of the final fuel. The control over the extent of these two competitive processes involved in the final refining of the hydrotreated alkane product is, however, a challenging task since the catalysts typically used for hydroisomerization also catalyze cracking reactions. [27,28]

A variety of feedstocks, reaction conditions and catalysts are currently employed for hydrotreating of lipids (Table 1). With regard to catalysts, two main classes of materials are used: (i) noble metals supported on classical carriers such as carbon and alumina and; (ii) metal sulfides supported on alumina. While the former have shown high deoxygenation activity for simple fatty acid model feeds, the latter are well-developed materials widely investigated for sulfur (HDS) and nitrogen (HDN) removal in the

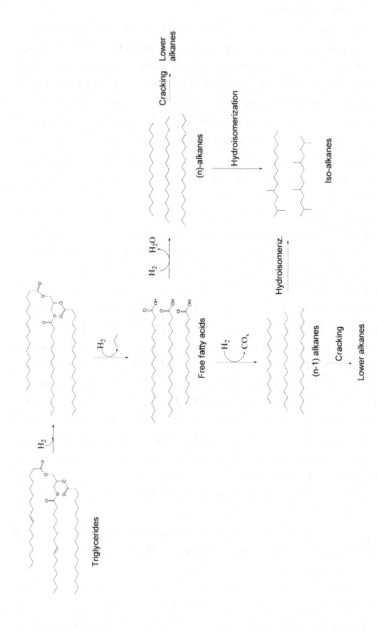

FIGURE 3: Scheme of the main reaction pathways involved in the hydrotreating of TGs to green jet fuel and diesel hydrocarbons.

petrochemical industry which also offer excellent performances for de-oxygenation of real feedstocks such as vegetable and waste oils.

Pt and, especially, Pd are the preferred noble metals for conversion of fatty acids into alkanes. Pd was found to be the most active element among a large variety of metal-based catalysts in the HDC of stearic acid to n-heptadecane, with the activity following the trend Pd > Pt > Ni > Rh > Ir > Ru > Os. [32] This trend was recently confirmed by Madsen et al. in the hydrotreating of oleic acid : tripalmitin mixtures to n-alkanes over alumina-supported Pt, Pd and Ni catalysts.35 Interestingly, Pd and Pt favored production of n-alkanes with odd number of carbon atoms (HDC products) versus Ni which showed higher HDO activity. Approaches based on the addition of a second metal have also been explored in an attempt to improve hydrotreating properties of monometallic catalysts. The performances of Ni in the processing of rapeseed-derived biodiesel to C15–C18 n-alkanes were enhanced upon addition of Cu which prevented undesirable hydrogen-consuming methanation ($CO_x + H_2 \rightarrow CH_4$) and cracking reactions. [36] Addition of rhenium in large amounts (20 wt%) allowed a dramatic increase of the conversion to hydrocarbons (from 15 to 80%) and a drastic decrease of the cracking activity of a Pt/H-ZSM5 catalyst used for the hydrotreating of Jatropha oil. [47] Re, in mixture with Pt, had been previously reported to offer outstanding activity for the deoxygenation of biomass sugars, [53] and this effect could be ascribed to the presence of ReO_x entities that assist in the C–O hydrogenolysis of oxygenated hydrocarbons associated with neighboring Pt sites. [54] The high cost of Re is, however, an important limitation for the commercial implementation of this hydrotreating technology.

Alumina-supported metal sulfides are, by far, the most used catalysts for the conversion of vegetable oils and related feedstocks into green hydrocarbons (Table 1). When compared with noble metal-based catalysts, metal sulfides show some important advantages such as lower cost, higher resistance to impurities typically present in waste feedstocks, and the possibility of using them in the co-hydroprocessing of plant lipids and petroleum feeds in existing refinery facilities. On the other hand, metal sulfides are more sensitive to the presence of water [55] (formed by HDO reactions), require a previous treatment with a sulfonation agent to reach the active state, and could potentially show sulfur leaching episodes leading

TABLE 1: Summary of recent technologies for the hydrotreating of triglycerides and related feedstocks to liquid hydrocarbon fuels[a]

Feedstock	Catalyst	Reaction conditions	Reference
Octanoic acid	Pd(3%)/C, Ni–Mo/Al$_2$O$_3$	300–400°C, 21 bar H$_2$/He, WHSV: 6 h^{-1}, H$_2$/oil: 1–20, CFBR	29
Dodecanoic acid	Pd(1%)/C	300–360°C, 5–20 bar Ar, WHSV: 1.7 h^{-1}, CFBR	30
Stearic acid	Pd(4%)/C	270–300°C, 17 bar He, oil/cat: 2.8, solvent: dodecane, [oil]: 0.05 M, SBR	31
Stearic acid	A variety of metal supported catalysts	300°C, 6 bar He, oil/cat: 4.5, solvent: dodecane, [oil]: 0.14 M, SBR	32
Methyl stearate	Pd(5%)/BaSO$_4$	270°C, 16 bar H$_2$, oil/cat: 1.9, solvent: hexane, [oil]: 0.01 M, BR	33
Stearic acid/ oleic acid	Pd(5%)/C	300°C, 15 bar He or 10% H$_2$/He, oil/cat: 4.6, solvent: dodecane, [oil]: 0.006 M, SBR	34
Oleic acid : tripalmitin (1 : 3)	Pt, Pd, Ni(5%)/Al$_2$O$_3$	325°C, 20 bar H$_2$, oil/cat: 4.5, solvent: tetradecane, [oil]b: 0.18 M, BR	35
FAME	Rh(0.5%)–Co, Ni(38%)–Cu, supported on various oxides	250–400°C, 5–20 bar H$_2$/Ar (50/50), LHSV: 1–6 h^{-1}, H$_2$/oil: 0.7–45, CFBR	36
Sunflower oil	Commercial supported metal sulfides	360–420°C, 180 bar H$_2$, WHSV: 0.7 h^{-1}, H$_2$/oil: 13, CFBR	37
Palm oil	NiMo/Al$_2$O$_3$	344°C, 40–90 bar H$_2$, WHSV: 0.7 h^{-1}, H$_2$/oil: 20, CFBR	38
Tristearin, tri-olein, soybean oil	Ni(20%)/C, Pd(5%)/C, Pt(1%)/C	350°C, 7 bar N$_2$, oil/cat: 91, no solvent, BR	39
Rapeseed oil, waste oil and trap grease	Sulfided Co–Mo/Al$_2$O$_3$	310°C, 35 bar H$_2$, WHSV: 2 h^{-1}, H2/oil: 100, CFBR	40
Fresh and used cooking oils	Sulfided hydrocracking commercial catalyst	350–390°C, 14 bar H$_2$, LHSV: 1.5 h^{-1}, H$_2$/oil: 15, CFBR	41
Sunflower oil	Sulfided NiMo/Al2O3/F	350–370°C, 20–40 bar H$_2$, LHSV: 1 h^{-1}, H$_2$/oil: 6–8, CFBR	42
Waste cooking oils and trap greases	Sulfided NiMo/Al$_2$O$_3$, NiW/Al$_2$O$_3$ and CoMo/Al$_2$O$_3$	350°C, 50 bar H$_2$, WHSV: 2.8 h^{-1}, H$_2$/oil: 9, CFBR	43
Sunflower oil	Pd(1%)/SAPO	310–360°C, 20 bar H$_2$, WHSV: 0.9–1.6 h^{-1}, H$_2$/oil: 14, CFBR	44
Rapeseed oil	Commercial NiMo/Al$_2$O$_3$	310–360°C, 70–150 bar H$_2$, WHSV: 1 h^{-1}, H$_2$/oil: 14, CFBR	45
Rapeseed oil	Sulfided NiMo/Al$_2$O$_3$	260–280°C, 35 bar H$_2$, WHSV: 0.25–4 h^{-1}, H$_2$/oil: 50, CFBR	46

TABLE 1: CONTINUED.

Jatropha oil	Pt(1%)–Re(10–20%)/ HZSM5	270–300°C, 65 bar H_2/N_2 (9/1), oil/ cat: 10, no solvent, BR	47
Rapeseed oil	Pt(2%)/HY, Pt(2%)/ HZSM5, Sulfided NiMo/ Al_2O_3	350°C, 50–110 bar H_2, oil/cat: 33, no solvent, BR	48
Rapeseed oil	Sulfided CoMo/Al_2O_3	250–350°C, 7–70 bar H_2, LHSV: 1.5 h^{-1}, H_2/oil: 100, CFBR	49
Sunflower oil in HVO	Sulfided NiMo/Al_2O_3	350–450°C, 50 bar H_2, WHSV: 5.0 h^{-1}, H_2/oil: 23, CFBR	50
Jatropha oil in gas oil mixtures	Sulfided NiW/SiO2– Al_2O_3, CoMo/Al_2O_3, NiMo/Al_2O_3	360°C, 50 bar H_2, LHSV: 1–2 h^{-1}, H_2/ oil: 21, CFBR	51
Rape oil in light gas oil mixtures	Sulfided NiMo/Al_2O_3	350–380°C, 50 bar H_2, LHSV: 2 h^{-1}, H_2/oil: 7, CFBR	52

a HVO: high vacuum petroleum oil; WHSV: weight hourly space velocity; LHSV: liquid hourly space velocity; CFBR: continuous fixed bed reactor; SBR: semi batch reactor, BR: batch reactor; H2/oil indicates molar ratio of H2 and oil in the feed (molar mass of oil taken as the main fatty acid component); [oil]: indicates molar concentration of the oil in the liquid solvent.; Oil/cat indicates oil to catalyst weight ratio; SAPO: silicoaluminophosphate. b Based on tripalmitin.

to contamination of the hydrotreated alkane mixture. [43] A comparative study of Pd/C and non-sulfided commercial Ni–Mo/Al_2O_3 in the hydrotreating of octanoic acid showed similar conversion levels to hydrocarbons for both materials. [29] However, while the Pd-based catalyst afforded almost exclusively the HDC product (e.g. n-heptane), Ni–Mo/ Al_2O_3 favored HDO to octane. This result has important implications since, as it will be remarked below, the HDC/HDO ratio is an important parameter determining the hydrogen consumption during hydrotreating processes.

As seen in Table 1, a variety of sulfided metals including Co, Mo, Ni and W (and combinations of them) are employed in the hydrotreating of vegetable oils. The performances of aluminasupported Ni, Mo and Ni–Mo catalysts in the deoxygenation of rapeseed oil were recently compared. [46] Both the activity and the selectivity to hydrocarbons were significantly improved for bimetallic Ni–Mo compared to monometallic catalysts, indicating a synergistic effect between both metals. In the same line, Ni–Mo and Ni–W based catalysts showed excellent yields to diesel-range alkanes and stability in the hydrotreating of waste [43] and Jatropha [51] oils. Both studies reported poorer performances (e.g. deactivation issues and high cracking activity) for Co–Mo/ Al_2O_3. This undesirable behavior was, however, not observed by Kubicka and coworkers. [49] The favorable conditions in terms of H_2 to oil molar ratio employed by these authors to perform their studies (H_2/oil : 100, Table 1) could have masked the intrinsic activity of Co–Mo in the hydroprocessing of rapeseed oil.

The support plays an important role in the conversion of vegetable oils to green hydrocarbons. Two characteristics need to be carefully controlled: porosity and acidity. The utilization of bulky TG molecules as feedstocks requires large pore size catalysts in order to avoid diffusion limitations. In this sense, it is estimated that the maximum dimension of a model TG such as triolein can be as high as 4.4 nm. [56] Consequently, supports with a large volume of micropores (pore size < 2 nm) are not suitable for hydrotreating of vegetable oils. Instead, materials with larger pore diameters (e.g. in the mesopore range) are preferred to ensure proper diffusion of reactant and products and to minimize potential pore blockage by coking and/or formation of heavier waxy hydrocarbons.

Support acidity is an important parameter determining the low temperature properties of the final fuel product. Thus, typical fuels obtained by hydrotreating of vegetable oils over non-acidic carbon or alumina supports consist of straight chain paraffins with very high cetane numbers (>85) but unfavorable pour and cloud points. Consequently, additional refining steps (e.g. isomerization) are normally required to enhance the cold flow properties of the hydrotreated fuel. [57] One alternative that is gaining interest in recent years involves utilization of bifunctional metal catalysts supported on acidic materials such as SiO_2–Al_2O_3, [51] fluorinated alumina, [42] B_2O_3–Al_2O_3, [43] silicoaluminophosphates (SAPO) [44] and zeolites. [48] This attractive approach allows deoxygenation and isomerization to take place in a single reactor thereby reducing complexity and cost in hydrotreating processes. However, the acidic strength of support materials needs to be moderate to maximize isomerization activity while keeping cracking at reasonable levels.

A large number of feedstocks including fatty acids and their methyl esters, TGs, edible and non-edible vegetable oils, waste cooking oils and trapped greases, animal fats, and mixtures of vegetable/petroleum oils are currently utilized for producing green hydrocarbons by diverse hydrotreating technologies (Table 1). Since HDC and HDO are the main reactions occurring in the hydrotreating reactor, the final composition of the alkane mixture can be roughly anticipated based on the fatty acid composition of the oil feed. As shown in Fig. 4a, the most representative vegetable oils and animal fats are essentially composed of C_{16} and C_{18} fatty acids and, consequently, the typical alkane product distribution obtained after hydroprocessing of these feedstocks ranges from C_{15} to C_{18}. Apart from the carbon chain length, the degree of unsaturations in the vegetable oil has important implications on the hydrotreating process as well. First, unsaturated fatty acids are more prone to suffering cracking at typical hydrotreating conditions and, consequently, vegetable oils with an elevated degree of unsaturations will produce higher amounts of light gaseous alkanes. [39] Second, a higher number of C=C bonds in the tri-chain structure will increase the hydrogen consumption during hydrotreating. Keeping these two aspects in mind, it is not strange that highly saturated palm oil and animal fats such as tallow (Fig. 4b) are preferred feedstocks for commercial hydrotreating production of green diesel hydrocarbons. [58] However, as rightly indicated by Knothe, [59] a number of unsaturations in the fatty acid chains is useful for the production of more branched hydrocarbons with better low-temperature properties such as those required for aviation. Thus, highly unsaturated feeds such as camelina [60] and algae-derived [61] oils are specially indicated for the production of hydrotreated jet fuels (HJF).

One of the main issues for the implementation of large-scale hydrotreating technologies is related to the availability and cost of the vegetable oil feed. In this sense, the utilization of waste and residues derived from existing industrial processes can help to improve the economics by offering cheaper (or even free of charge) additional feedstock alternatives to limited virgin vegetable oils. Waste cooking oils and trapped greases, abattoir fat wastes, and tall oil (a by-product of Kraft pulping of pine rich in fatty acids) are good examples in this regard, and their annual production can reach considerable amounts. For example, U.S. restaurants alone produce around 1.2×10^{10} L of waste vegetable oil a year [62] that could potentially cover up to 4% of the diesel demand of this country if converted to equal amounts of fuel. [35] However, the hydrotreating of these waste

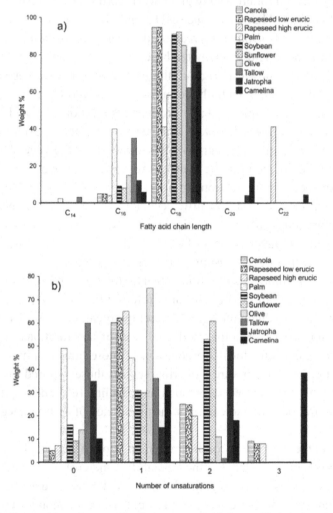

FIGURE 4: Composition of vegetable oils in terms of: (a) fatty acid chain length; and (b) the number of unsaturations in the carbon chain.

feedstocks is sometimes troublesome due to the presence of impurities that can seriously affect the performances of catalysts. In this sense, the use of metal-sulfides is recommended since these materials, unlike noble metal-based catalysts, can cope with typical impurities of waste feedstocks such as S and N facilitating their removal as H_2S and NH_3 while keeping acceptable hydrocarbon yields and stability versus time on stream. [41,43] These results contrast, however, with those reporting a strong inhibition effect of NH_3 over HDO and HDC of carboxylic acids. [63] More problematic is the presence of alkalis and phospholipids. The former deposit on the catalyst surface leading to blockage of active sites and/or plugging issues, and the latter release phosphoric acid that catalyzes coke-forming oligomerization reactions. [40]

The reaction conditions are important parameters controlling the nature of the final hydrotreated product. Thus, a minimum temperature of 300–350°C is typically required to achieve complete deoxygenation of the vegetable oil. This range of temperatures favors production of straight-chain alkanes versus isomerization and cracking products that are typically formed at harsher (e.g. 350–400°C) conditions. [37,48] Cold flow properties of hydrotreated fuel can thus be also enhanced by proper control of reaction temperature. [42,45] The HDC/HDO ratio is also controlled by temperature and pressure reaction conditions. Numerous studies confirmed that HDC is favored at higher temperatures and low pressures [42,44,46,50] while HDO prevailed at high hydrogen pressures. [38,42,48]

Hydrogen consumption is, along with the cost of the feed and the catalyst, the key parameter affecting the economics of any hydrotreating technology. It has been estimated that approximately 7–16 moles of H_2 per mol of TG are required for the hydrotreating of vegetable oils and related feedstocks. [26] Strategies to reduce hydrogen consumption during hydrotreating are highly necessary since this gas is expensive and typically derived from external fossil fuel sources. Factors such as the degree of unsaturations of the TGs in the vegetable oil, the HDC/HDO ratio, reaction conditions (e.g. H_2 partial pressure, H_2/oil feed ratio) and reactor type (batch versus flow reactors) are important parameters affecting hydrogen consumption, and all of them involve trade-offs. For example, the utilization of highly saturated feedstocks such as palm and tallow allows significant reduction of hydrogen consumption at the expense of having feeds less prone to undergo hydroisomerization. HDC reactions are preferred since they are considerably less hydrogen demanding [26] and only slightly penalized in terms of maximum hydrocarbon yield (81 vs. 86 wt%) compared to HDO. However, secondary hydrogen-consuming reactions involving COx species such as reverse water gas shift (RWGS,

$CO_2+H_2 \rightarrow CO + H_2O$) and methanation could reduce the attractiveness of the HDC route if these unwanted processes are not adequately controlled by a proper selection of reaction conditions and catalysts.

Hydrogen consumption increases exponentially with H_2 partial pressure in the hydrotreating reactor. [38] This fact has motivated researchers to explore the possibility of performing hydrotreating under inert (e.g. N_2, He) atmospheres [30,34,39,64] although the severe deactivations reported [30,33,64] suggest that a minimum amount of hydrogen seems to be necessary to maintain activity and prevent formation of unsaturated intermediates that polymerize forming coke. A recent breakthrough was reported by Fu and coworkers [64,65] in which HDC of fatty acids to n-alkanes takes place with high selectivity (e.g. 90%), good stability and without H_2 requirements. The key of this promising technology lies in the utilization of near supercritical water as a reaction medium which provides the in situ H_2 necessary to perform hydrotreating cleanly.

Hydrotreated diesel and jet fuels are currently produced commercially by several two-stage (deoxygenation + isomerization) trademark processes such as NExBTL™(Neste oil) and Eni Ecofining™(UOP). Neste oil has currently three operative plants in Finland and Singapore with a production capacity of 1×10^6 metric tons of green diesel per year, [66] and a fourth plant in Rotterdam under construction that will double production by 2011. ENI announced construction of the first Ecofining unit in Italy operative in 2010 with a capacity of 2.5×10^5 tonnes per year. [67] Both NExBTL™ and Eni Ecofining™ synthetic diesel fuels show comparable properties to petroleum-derived ultra low sulfur diesel (ULSD) and superior stability, heating value and pollutant emission characteristics than conventional biodiesel (Table 2).

The stringent requirements of aviation fuels (e.g. high energy density and efficient use at very low temperatures) have placed HJF in a privileged situation versus biodiesel. Governments, through ambitious initiatives such as the "European Advanced Biofuels Flight Path" taken by the European commission, leading European airlines and biofuel producers are trying to accelerate commercialization of drop-in biofuels for aviation with ambitious targets to produce 2×10^6 tonnes of paraffinic jet fuel per year by 2020. [68] Hydrotreating technologies, along with biomass to liquids (BTL) and pyrolysis approaches, [11] will be crucial for the

TABLE 2: Comparison of properties of ULSD, biodiesel and green diesel derived from hydrotreating of vegetable oils. Adapted from ref. 78 and 114

	Petroleumderived ULSD	Biodiesel	Hydrotreated green diesel
Density/g mL^{-1}	0.84	0.88	0.78
Oxygen content (wt%)	0	11	0
Sulfur content/ppm	<10	<1	<1
Polyaromatics (wt%)	11	0	0
Cetane number	40	50–65	70–90
Heating value MJ kg^{-1}	43	38	44
Cloud point/C	−5	−5 to 15	−10 to 5
NO$_x$ emissions (versus ULSD)	—	+10%	−10%
CO$_2$ emissionsa/kg MJ^{-1}	0.08	0.06	0.04
Stability (versus ULSD)	—	Poor	Excellent

[a] Data taken from ref. 115.

accomplishment of these targets. Recently, the ASTM has approved a new standard that allows utilization of HJF in commercial flights and military jets blended with regular jet fuel in amounts up to 50% (v/v). [69] Aviation companies such as Lufthansa [70] and KLM [71] have recently announced commercial flights on 50/50 HJF/conventional jet fuel mixtures. Interestingly, the main impediment for the utilization of 100% pure HJF in current aircraft is the lack of aromatics (8% minimum in conventional jet fuels [72]) which are necessary to swell elastomeric valves in fuel systems, providing a proper seal and avoiding leaks. [73] Further research is thus needed to find a fuel additive that ensures elastomer swell even in the absence of aromatics.

Fig. 5 shows a comparative scheme of transesterification and hydrotreating processes. Both technologies utilize TGs as feedstocks but they differ in the reactants utilized (methanol vs. hydrogen), the by-products generated (glycerol vs. propane), the final fuel product obtained (biodiesel vs. green hydrocarbons) as well as in the reaction conditions and catalysts

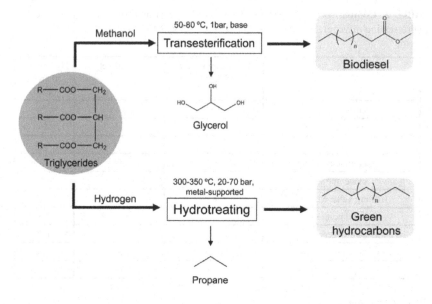

FIGURE 5: Comparative scheme of transesterification and hydrotreating processes for the conversion of TGs into biodiesel and green hydrocarbons, respectively.

used. Methanol and hydrogen are typically derived from fossil fuels and, consequently, efforts should be made to obtain these reactants from biomass sources in order to reduce the overall CO_2 footprint of the biofuel. While solutions in the biodiesel industry involve replacement of methanol with biomass-derived ethanol as an esterification agent, hydrotreating technologies can drastically reduce external hydrogen consumption by employing sub- products and/or residues generated during the process as sources of this gas. For example, up to 75% of the H_2 needs of hydrotreating can be covered by steam reforming and subsequent WGS of the propane co-produced during the process, [74] while the lignocellulosic soybean hull wastes discarded after oil extraction can provide hydrogen for hydroprocessing by means of microbial fermentation. [75] The higher cost of hydrogen compared to methanol should be a strong incentive to implement the mentioned solutions (or similar approaches) in commercial hydrotreating plants.

The separation and subsequent management of the by-products generated during the process is also an important aspect determining the profitability of both technologies. In this sense, transesterification seems to be more sensitive to this parameter given the large amounts of glycerol generated and the difficulty to completely remove it from the biodiesel fuel (Fig. 2). However, once separated, this crude glycerol can serve as a cheap feedstock for the production of a large variety of high value-added chemicals and fuels [76] thereby representing an opportunity to reduce overall biodiesel production costs. [77] Hydrotreating, on the other hand, generates a by-product gas stream enriched in propane which is easily separable from the liquid hydrocarbon fuel but presents a lower chemical value compared to glycerol. Consequently, rather than to decrease the costs, this gas stream could be important to reduce the overall input of fossil fuels in the process by offering an internal source of hydrogen or heat/ electricity.

Transesterification of lipid feedstocks requires milder temperature and pressure conditions compared to hydrotreating and thus operational costs are greater for the latter route. However, hydrotreating conditions are similar to those typically used in HDN and HDS of petroleum which opens the possibility to co-process lipids and fossil feeds in existing refinery facilities (Table 1). [50, 78] This synergy between hydrotreating and conventional oil refineries would greatly reduce capital costs, [79] and represents one of the key advantages of hydrotreating versus conventional transesterification. However, aspects such as the corrosion of the hydroprocessing reactor by free fatty acids, the detrimental cold flow properties of the diesel product as a consequence of the increased content of n-alkanes, [26, 52] and the effect of the presence of oxygenates over intrinsic HDNand HDS activities of commercial hydroprocessing catalysts are key points that still require further research studies.

The simplicity of the chemistry involved in transesterification and hydrotreating allows production of biodiesel and green hydrocarbons with high yields. In this sense, both technologies benefit from the utilization of feeds with relatively low oxygen content (and thus low reactivity) like TGs to achieve the required transformations in a selective fashion, and this represents an important advantage versus other biomass conversion routes managing more reactive feedstocks such as sugars or lignocellulosic biomass. The latter feedstocks are, however, more abundant and cheaper than vegetable oils. The limited availability of lipids to satisfy the growing demand for both biodiesel and green hydrocarbon fuels is, by far, the most important issue facing both transesterification and hydrotreating technologies. It is therefore imperative to search for additional and preferably non-edible sources of lipids that can ensure sustainable supply without affecting food markets or requiring large land extensions. Hydrotreating presents higher flexibility to cope with different kinds of feeds compared to transesterification, which is more sensitive to the presence of impurities or free fatty acids. In this sense, hydrotreating is better positioned for the implementation of new, more abundant and non food-competitive feedstocks such as algae [80] and lignocellulosic residues-based microbial [81] oils in near future.

6.3 MICROBIAL ROUTES TO LIQUID HYDROCARBON FUELS

The utilization of microbes of different classes (e.g. bacteria, yeast, fungi, algae) for the production of advanced biofuels has experienced tremendous progress in the last few years. The improvements in the so-called metabolic engineering (i.e. the genetic manipulation of a microbe to change its inherent metabolism favoring pathways to desired products); the vast knowledge acquired about the metabolic functioning and genetic information of model microbes such as *Escherichia coli* (*E. coli*) and *Saccharomyces cerevisiae* (*S. cerevisiae*); and the development of new useful analytical and computer-modeling tools have allowed the design of microbes for production of a variety of advanced biofuels including gasoline-compatible alcohols (C_4–C_7) and alkanes for diesel and jet fuel applications. However, some of these compounds are naturally produced by microbes as a part of their internal machinery. For example, it is well known that some bacteria, yeast and fungi naturally synthesize alkanes to build protection structures and as a mechanism of energy storage. [82] It is also believed that these alkaneproducing microorganisms could play a role in the petroleum formation process. [83] As will be described below, new biotechnologies take advantage of this natural ability by means of two different approaches: [84] (i) engineering the natural-producer microorganism to boost its advanced fuel production capability or (ii) extracting the genetic information leading to fuel production from an external microbe and inserting it in a wellknown and robust host.

Fig. 6 shows a scheme of the microbial reaction pathways involved in the production of fuels from sugars. Four different routes are available today for the microbial synthesis of fuels: fermentative, non-fermentative, isoprenoid, and fatty-acid pathways, [85] and in all of them metabolic engineering is currently playing a crucial role. Fermentation of hexoses to ethanol (Fig. 1) is, by far, the most studied and mature route. Although this technology is well developed for easily degradable starchy materials, the inclusion of more recalcitrant and heterogeneous feedstocks such as lignocellulose has added new difficulties to the process, and biotechnology is actively involved in solving these issues. For example, genetic manipulation of yeasts has allowed processing of C_5 sugars into ethanol, [86] whereas ordinary yeasts are unable to convert these important components of lignocellulose. Additionally, a new generation of yeasts with the ability to secrete enzymes such as cellulases and hemicellulases are being tested with the aim to consolidate the biomass depolymerization and fermentation steps in a single reaction unit. [87] Fermentation technologies are not constrained to ethanol. Recently, butanol, with polarity and energy-density properties that resemble those of gasoline, has attracted interest as an infrastructure-compatible biofuel, [88] and current biotechnology efforts are focussed on designing new organisms with boosted butanol-producing abilities and more resistance to accumulated products. [85]

Hydrophobic C_4–C_7 alcohols with superior fuel properties than ethanol as replacement of gasoline can be produced from sugars by a non-fermentative pathway. This non-naturally occurring route, developed by Liao and coworkers, [89, 90] is based on the formation of 2-ketoacids which are intermediates in the metabolic biosynthesis of aminoacids (Fig. 6). These researchers used the known ability of yeast *S. cerevisiae* to convert ketoacids into alcohols [91] and implanted it into *E. coli* allowing production of alcohols such as isobutanol, 1-butanol, 2-methyl-1-butanol, 3-methyl-1-butanol, and 2-phenylethanol at concentrations in the order of mg L^{-1}. Importantly, by overexpression of the genes responsible for the amino acid synthesis and by deletion of those involved in competing pathways these authors achieved production of isobutanol at outstanding concentrations of 22 g L^{-1} which represents 86% of the theoretical maximum yield for the glucose to isobutanol process. [89] Furthermore, owing to its low solubility in water, isobutanol separates from the aqueous broth thereby avoiding additional energy-consuming distillation steps. This fact, along with the potential utilization of existing ethanol fermenters to carry out the isobutanol microbial synthesis, gives this process a good economic perspective. Gevo, a renewable chemicals and advanced biofuels company, [92] has licensed the non-fermentative isobutanol technology for the commercial production of, among other interesting products, green gasoline and jet fuels. The process involves well-known chemistry and existing refinery infrastructure. First, isobutanol is dehydrated over a solid acid catalyst to isobutene which is subsequently oligomerized to produce branched aliphatic oligomers such as isooctane (valuable for gasoline) or C_{12} jet fuel components. Interestingly, a fraction of isobutanol is utilized to produce aromatics which can be added to aliphatic components to create a fuel virtually identical to those currently used in aviation. This represents an important advantage versus other fully aliphatic jet fuels (such as HJF described in Section 2) that opens the possibility of using 100% renewable kerosene in the near future.

Isoprenoids are a large family of hydrocarbon compounds derived from the C_5 monomer unit isoprene (2-methyl-1,3- butadiene) which are synthesized naturally in plants, animals and bacteria. Although the isoprenoid route was classically exploited to synthesize compounds with pharmaceutical [93] and nutritional value, it has a great potential for the production of hydrocarbon fuels. Microorganisms convert glucose into isoprenoids by two different routes which lead to two isomeric C_5 isoprenoid units denominated as isopentenyl pyrophosphate (IPP) and dimethylallyl pyrophosphate (DMAPP) (Fig. 6). A more detailed description of this important metabolic route can be found in excellent reviews by Keasling's group. [94, 95] IPP and DMAPP monomers are subsequently polymerized to form larger C_{10}, C_{15} and C_{20} activated units which represent the base for the production of isoprenoid fuels. These units can be recombined in multiple ways to yield a large variety of compounds with different molecular weights and levels of branching suitable for gasoline, diesel and jet fuel applications. In this sense, farnesene, a C_{15} branched olefin with potential to serve as a winter-diesel component after hydrogenation, is produced in genetically modified *E. coli* and *S. cerevisiae* with yields exceeding 14 g

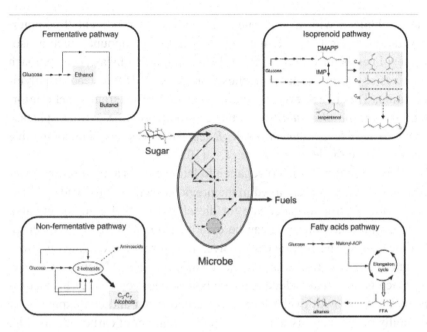

FIGURE 6: Scheme of the reaction pathways available today for the microbial production of fuels. Adapted from ref. 85 and 94.

L^{-1}. [96] This technology is currently exploited by Amyris Biotechnologies Inc. [97] to produce a spontaneously separating from the broth [98] renewable diesel precursor (Biofene) from sugarcane-derived glucose. The fuel derived from Biofene has been recently allowed by the US EPA to be blended with petroleum-derived ULSD in amounts up to 35% (v/v), [99] and the company plans to produce 1×10^9 L of farnesene by 2015. [100] Amyris is also exploring the introduction of external terpene synthase enzymes into *E. coli* and *S. cerevisiae* for the production of cyclic C_{10} terpenes such as pinene, sabinene or terpinene which are proposed as a new generation of jet fuel components. [101]

Microorganisms can be genetically manipulated to induce metabolic polymerization of isoprene units to high molecular weight compounds ($C_{>30}$) generating a petroleum-like mixture denoted as microbial biocrude. This technology is promising in that biocrude could be further processed to fuels and chemicals in existing petroleum refineries. However, the low yields obtained with *E. coli* and *S. cerevisiae* have prevented commercialization so far. [85] Alternatively, lighter fuel compounds such as isopentenol can be also derived from the isoprenoid pathway. The approach involves re-routing to favor dephosphorylation of IPP instead of polymerization. This is achieved by the introduction of external genetic information from *Bacilus subtilis* (leading to production of a pyrophosphatase enzyme) into *E. coli*. As a result, *E. coli* achieves gasoline-compatible isopentenol to the level of 112 mg L^{-1}. [102]

Many organisms convert sugars into fatty acids as a natural way to accumulate energy. As remarked in the previous section, fatty acids possess the chemical structure adequate for diesel and jet fuel components and, consequently, this bioroute can be exploited for the generation of green hydrocarbons. The bioprocess involves transformation of glucose into a C_2 intermediate carrier (malonyl acyl carrier protein, Malonyl-ACP, Fig. 6) which is repeatedly added to a metabolic elongation cycle. This mechanism allows growth of the fatty acid carbon chain and can explain why naturally occurring fatty acids have an even number of carbon atoms (Fig. 4a). When the fatty acid reaches a determined chain length it is liberated from the elongation cycle and incorporated to the cell membrane. New biotechnologies pursue modification of this natural pathway to avoid fatty acids storage thereby favoring subsequent reactions to fuels. Remarkably, genetic manipulation allows control over these processes to generate fuels of diverse classes. For example, by expressing external genes involved in the production of TGs and in the biosynthesis of ethanol it is possible to bioproduce fatty acid ethyl esters (FAEE) in *E. coli*. [103] Unlike the conventional biodiesel process which is limited to a determined composition fixed by the vegetable oil feed, this technology allows fine manipulation of metabolic routes to produce FAEE with specific structures (e.g. different carbon chain lengths, levels of unsaturations and branching, etc.) that better fulfil stringent requirements of diesel fuels. [104] *E. coli* microbes can even be manipulated for simultaneous secretion of hemicellulases to assist in the depolymerization of hemicellulose within the same biocatalytic reactor. [105] This promising technology opens the possibility of producing biodiesel from cheaper and more abundant lignocellulose- derived sugars.

Perhaps the most interesting process involving the fatty acid route is the production of aliphatic straight-chain hydrocarbons. Although the metabolism for this pathway is not well understood, it is believed that fatty acids undergo reduction to the corresponding aldehydes followed by decarbonylation to yield a (n − 1) linear alkane. [85] The fact that most of the alkanes found in biological systems have an odd number of carbon atoms supports this hypothesis. A recent breakthrough on microbial production of linear hydrocarbons was reported by Schirmer and coworkers.106 These authors identified two key genes involved in the natural hydrocarbon synthesis metabolism of some cianobacteria.

These genes were responsible for the creation of reductase and decarbonylase enzymes involved in the fatty acid reduction and aldehyde decarbonylation processes, respectively. Expression of these pieces of genetic information in *E. coli* allowed secretion of a mixture of tridecane (C_{13}), pentadecene ($C_{15=}$), pentadecane (C_{15}) and heptadecene ($C_{17=}$) in a ratio of 10 : 10 : 40 : 40 and at concentrations of 300 mg L^{-1}. The same researchers have recently identified genes involved in the synthesis of C20 terminal olefins in another group of bacteria and successfully expressed them in *E. coli*. [107] LS9 Inc. [108] is currently exploiting these technologies for the microbial production of aromatics-free renewable UltraClean™ diesel from sugars. As in the case of isoprenoid-based fuels, this route benefits from spontaneous separation of the hydrocarbons from the aqueous broth [109] making purification simple and reducing fuel costs. Since the US EPA approval in April 2010, LS9 UltraClean diesel can be sold commercially in the US. [110]

Yeasts and bacteria such as *E. coli* are preferred as host microorganisms since they are genetically well-known and easily manipulable systems. Photosynthetic algae or bacteria are, on the other hand, more complex and difficult to decode. However, these organisms are attractive since, unlike sugar-consuming microbes, they have the ability to generate hydrocarbons with the only input of sunlight, water and CO_2 (Fig. 7). This represents an enormous advantage versus conventional sugar-based routes since it avoids utilization of biomass and the subsequent expensive lignocellulose pre-treatment and hydrolysis steps required for the generation of the sugar feed. The direct photosynthetic production of alkanes utilizes a readily transformable cianobacterium which is genetically modified to provide this microorganism with the ability to produce linear alkanes, to efficiently secrete them outside the cell, and to allow equilibrated partitioning of the carbon fed between alkane production and cell growth metabolic routes. [111] Joule Unlimited Technologies [112] has patented this promising process denominated as Helioculture in which alkanes are continuously produced from microorganisms located in solar converters. The process, which has the capacity to generate almost 6×10^4 L of photosynthetic alkanes per acre (1 acre ≈ 4000 m^2) and per year, can be fed with non- potable water and industrial waste CO_2 at concentrations 50–100 times higher than atmospheric, and represents an improvement over indirect photosynthetic algae approaches requiring growing, extracting and hydrotreating conversion of TGs to produce the same final alkane product (Fig. 7).

6.4 FUTURE PROSPECTS AND CONCLUSIONS

Advanced biofuels such as green hydrocarbons represent an attractive alternative to conventional bioethanol and biodiesel. The conversion of biomass into gasoline, diesel and jet fuels is a paradigmatic process that would help to overcome many of the current compatibility and energy-density issues associated with oxygenated biofuels. The large incentives of green hydrocarbons are favoring the onset of a new group of promising technologies which, in many cases, are based on previously well-known processes. Hydrotreating (petroleum hydroprocessing) and microbial syntheses (classical fermentation) are among this group of technologies that have taken advantage of previous mature technologies to develop fast and reach the commercial scale. Conceptually, these two technologies can be considered as the modified versions of classical transesterification and fermentation with the aim of producing hydrocarbons instead of oxygenates. Thus, while in hydrotreating the process is adapted by employing different catalysts, reactants and harsher conditions to allow complete deoxygenation of TGs, in the case of microbial synthesis the modification is carried out inside microorganisms to favor sugar metabolic pathways leading to green hydrocarbons instead of ethanol.

Hydrotreating of vegetable oils has demonstrated great potential for high yield production of fungible diesel and jet fuels. However, some important challenges lie ahead. New strategies for reducing external hydrogen consumption and additional tests in existing refinery facilities are

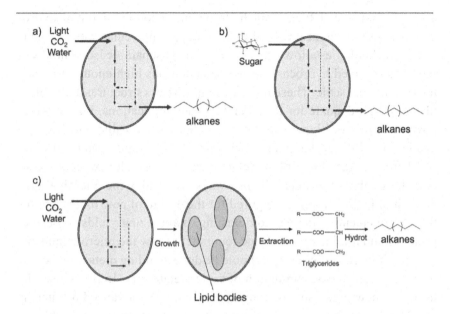

FIGURE 7: Schematic comparison for the different approaches utilized today for the microbial production of alkanes. (a) Direct process for genetically modified photosynthetic cianobacterium; (b) direct process for sugar-consuming microbes (heterotrophic) such as bacteria, yeast and fungi; and (c) process for algae transformation into liquid hydrocarbon fuels via extraction of TGs and subsequent hydrotreating. Adapted from ref. 111.

required to decrease capital expenditures. Taking into account the high cost of vegetable oils compared to lignocellulose, this is a crucial point to ensure the economic feasibility of this route and the competitiveness with other hydrocarbons-producing emerging technologies such as BTL and pyrolysis. The efficient utilization of new non-edible oil feedstocks such as those derived from algae and microbes or industrial residues enriched in lipids is crucial to alleviate the availability issue that seriously jeopardizes the future expansion of the hydrotreating industry. The design of new robust catalysts able to cope with the impurities typically present in these waste feedstocks will help accomplish this task.

The science of genetic manipulation of microorganisms has reached the point in which a designer microbe can be tailored for production of a specific advanced fuel, from hydrophobic alcohols to liquid hydro-carbons with different levels of branching and molecular weights. Current biotechnologies allow production of easily separable isoprenoid and fatty acid-derived hydrocarbons at concentrations high enough to reach the commercial scale. These hydrocarbons possess structures and molecular weights adequate for diesel and jet fuel applications. Furthermore, a whole family of higher alcohols, potentially convertible into gasoline and jet fuels by well-known petrochemical technologies, can be also derived from sugars by nonfermentative approaches. However, a deeper knowledge about the metabolic pathways involved in the microbial production of hydrocarbons is required for the design of new routes and for the improvement of the already existing, maximizing yields of desired products and minimizing the time required to grow the microorganisms. The manipulation of microbes should be done avoiding metabolic bottle-necks or excessive accumulation of toxic metabolites. In this sense, the design of new organisms with higher resistance to advanced biofuels is crucial. This can be done, for example, by isolating genes responsible for efflux pumps (a natural mechanism of microbes to remove impurities out of the cell) and inserting them into *E. coli*. [113] Furthermore, the design of new cost-effective enzymes for deconstruction of lignocellulose into transformable sugars is crucial for the success of this route which currently relies on edible biomass sources such as sugar cane. Direct photosynthetic technologies, able to synthesize alkanes with no carbon input other than CO_2, are considered excellent alternatives to conventional sugar-based microbial approaches.

REFERENCES

1. Intergovernmental panel on climate change, Climate Change 2007: Synthesis Report, http://www.ipcc.ch/publications_and_data/ar4/ syr/en/contents.html, accessed August 2011.
2. American Energy: the Renewable Path to Energy Security, Worldwatch Institute Center for American Progress, 2006, p. 19, http://images1.americanprogress.org/il80web20037/americanenergynow/AmericanEnergy.pdf, accessed August 2011.
3. Bionomicfuel.com, Fossil Fuels Reserves Will Not Last Long, http://www.bionomicfuel.com/fossil-fuels-reserves-will-not-last-long-full, accessed August 2011.
4. A. J. Ragauskas, et al., Science, 2006, 311, 484.

5. US Energy Information Administration (EIA), International Energy Outlook, 2010, http://www.eia.doe.gov/oiaf/ieo, accessed August 2011.
6. R. Luque, L. Herrero-Davila, J. M. Campelo, J. H. Clark, J. M. Hidalgo, D. Luna, J. M. Marinas and A. A. Romero, Energy Environ. Sci., 2008, 1, 542.
7. D. Martin-Alonso, J. Q. Bond and J. A. Dumesic, Green Chem., 2010, 12, 1493.
8. J. P. Lange, Biofuels, Bioprod. Biorefin., 2007, 1, 39.
9. H. Shapouri, J. A. Duffield and M. Wang, in The Energy Balance of Corn Ethanol: an Update (report no. 814, Office of the Chief Economist), U.S. Department of Agriculture, 2002, http://www.transportation.anl.gov/pdfs/AF/265.pdf, accessed August 2011.
10. U.S. National Science Foundation, in Breaking the Chemical and Engineering Barriers to Lignocellulosic Biofuels: Next Generation Hydrocarbon Biorefineries, 2008, http://www.ecs.umass.edu/biofuels/Images/Roadmap2-08.pdf, accessed August 2011.
11. J. C. Serrano-Ruiz andJ. A.Dumesic, EnergyEnviron. Sci., 2011, 4, 83.
12. Public Law 110–140, Energy Independence and Security Act of 2007, 2007, http://energy.senate.gov/public/_files/getdoc1.pdf, accessed August 2011.
13. R. F. Service, Is there a road ahead for cellulosic ethanol?, Science, 329, 784.
14. W. E. Tyner, F. J. Dooley and D. Vitery, Am. J. Agr. Econ., 2010, 93, 465.
15. EPA/DOE sponsored web site, http://www.fueleconomy.gov/feg/flextech.shtml, accessed August 2011.
16. U.S. Environmental Protection Agency report, Water Phase Separation in Oxygenated Gasoline, http://www.epa.gov/oms/regs/fuels/rfg/waterphs.pdf, accessed August 2011.
17. Biodiesel Magazine, Report: 12 billion gallons of biodiesel by 2020, March 2010, http://www.biodieselmagazine.com/article.jsp?article_id ¼ 4080, accessed August 2011.
18. J. M. Marchetti, V. U. Miguel and A. F. Errazu, Renewable Sustainable Energy Rev., 2007, 11, 1300.
19. S. Al-Zuhair, Biofuels, Bioprod. Biorefin., 2007, 1, 57.
20. R. Luque, Energy Environ. Sci., 2010, 3, 254.
21. Worldwatch Institute, Biofuels for Transport, Earthscan, London, 2007.
22. J. R. Regalbuto, Biofuels, Bioprod. Biorefin., 2011, 5, 495.
23. I. Kubickova and D. Kubicka, Waste Biomass Valoriz., 2010, 1, 293.
24. J. R. Regalbuto, Science, 2009, 325, 822.
25. N. Savage, Nature, 2011, 474, S9.
26. B. Donnis, R. G. Egeberg, P. Blom and K. G. Knudsen, Top. Catal., 2009, 52, 229.
27. V.Calemma, S. Peratello andC. Perego,Appl.Catal.,A, 2000, 190, 207.
28. Y. Liu, R. Sotelo-Boyas, K. Morata, T. Minowa and K. Sakanishi, Chem. Lett., 2009, 28, 552.
29. L. Boda, G. Onyestyak, H. Solt, F. Lonyi, J. Valyon and A. Thernesz, Appl. Catal., A, 2010, 374, 158.
30. H. Bernas, et al., Fuel, 2010, 89, 2033.
31. S. Lestari, I. Simakova, A. Tokarev, P.M€aki-Arvela, K. Er€anen and D. Y. Murzin, Catal. Lett., 2008, 122, 247.

32. M. Snare, I. Kubikova, P. M€aki-Arvela, K. Er€anen and D. Y. Murzin, Ind. Eng. Chem. Res., 2006, 45, 5708.
33. J. Han, H. Sun, Y. Ding, H. Lou and X. Zheng, Green Chem., 2010, 12, 463.
34. J. G. Immer,M.J. Kelly and H. H. Lamb, Appl.Catal.,A, 2010, 375, 134.
35. A. T. Madsen, E. H. Ahmed, C. H. Christensen, R. Fehrmann and A. Riisager, Fuel, 2011, 90, 3433.
36. V. A. Yakovlev, et al., Catal. Today, 2009, 144, 362.
37. P. Simacek, D. Kubicka, I. Kubickova, F. Homola, M. Pospisil and J. Chudoba, Fuel, 2011, 90, 2473.
38. A. Guzman, J. E. Torres, L. P. Prada and M. L. Nu~nez, Catal. Today, 2010, 156, 38.
39. T. Morgan, D. Grubb, E. Santillan-Jimenez and M. Crocker, Top. Catal., 2010, 53, 820.
40. D. Kubicka and J. Horacek, Appl. Catal., A, 2011, 394, 9.
41. S. Bezergianni, S. Voutetakis and A. Kalogianni, Ind. Eng. Chem. Res., 2009, 48, 8402.
42. S. Kovacs, T. Kasza, A. Thernesz, I. W. Horvath and J. Hancsok, Chem. Eng. J., 2011, 176-177, 237, DOI: 10.1016/j.cej.2011.05. 110.
43. M. Toba, Y. Abe, H. Kuramochi, M. Osako, T. Mochizuki and Y. Yoshimura, Catal. Today, 2011, 164, 533.
44. O. V. Kikhtyanin, A. E. Rubanov, A. B. Ayupov and G. V. Echevsky, Fuel, 2010, 89, 3085.
45. P. Simacek, D. Kubicka, G. Sebor and M. Pospisil, Fuel, 2010, 89, 611.
46. D. Kubicka and L. Kaluza, Appl. Catal., A, 2010, 372, 199.
47. K. Murata, Y. Liu, M. Inaba and I. Takahara, Energy Fuels, 2010, 24, 2404.
48. R. Sotelo-Boyas, Y. Liu and T. Minowa, Ind. Eng. Chem. Res., 2011, 50, 2791.
49. D. Kubicka, P. Simacek and N. Zilkova, Top. Catal., 2009, 52, 161.
50. G. W. Huber, P. O'Connor and A. Corma, Appl. Catal., A, 2007, 329, 120.
51. R. Kumar, et al., Green Chem., 2010, 12, 2232.
52. J. Walendziewski, M. Stolarski, R. Luzny and B. Klimek, Fuel Process. Technol., 2009, 90, 686.
53. E. L. Kunkes, D. A. Simonetti, R. M. West, J. C. Serrano-Ruiz, C. A. G€artner and J. A. Dumesic, Science, 2008, 322, 417.
54. E. L. Kunkes, et al., J. Catal., 2008, 260, 164.
55. O. I. Senol, T. R. Viljava and A. O. I. Krause, Catal. Today, 2005, 106, 186.
56. T. J. Benson, R. Hernandez, W. T. French, E. G. Alley and W. E. Holmes, J. Mol. Catal. A: Chem., 2009, 303, 117.
57. J. Hancsok, M. Krar, S. Magyar, L. Boda, A. Hollo and D. Kallo, Microporous Mesoporous Mater., 2007, 101, 148.
58. http://www.biodieselmagazine.com/articles/3499/neste-oil-starts-construction-on-europe%27s-largest-renewable-fuels-plant, accessed August 2011.
59. G. Knothe, Prog. Energy Combust. Sci., 2010, 36, 364.
60. Biofuels taking flight, Seattle Business Magazine, http://climatesolutions.org/programs/aviation-biofuels-initiative/press-releases/biofuels-taking-fligh/at_download/file, accessed August 2011.
61. Biodiesel magazine, Boeing planes successfully fly with biofuels, http://biodieselmagazine.com/articles/3141/boeing-planes-succesfully-fly-withbiofuels, accessed August 2011.

62. www.OilrecyclingWNY.com, accessed August 2011.
63. E. Laurent and B. Delmon, Appl. Catal., A, 1994, 109, 97.
64. J. Fu, X. Lu and P. E. Savage, ChemSusChem, 2011, 4, 481.
65. J. Fu, X. Lu and P. E. Savage, Energy Environ. Sci., 2010, 3, 311.
66. http://www.nesteoil.com/default.asp?path ¼ 1,41, 11991, 12243, 12335, 12337, accessed August 2011.
67. http://www.uop.com/processing-solutions/biofuels/green-diesel, accessed August 2011.
68. http://ec.europa.eu/energy/technology/initiatives/biofuels_flight_path_en.htm, accessed August 2011.
69. ASTM D7566-11 Standard Specification for Aviation Turbine Fuel Containing Synthesized Hydrocarbons, DOI: 10.1520/D7566-11.
70. http://presse.lufthansa.com/en/news-releases/singleview/archive/2010/november/29/article/1828.html, accessed August 2011.
71. http://www.klm.com/corporate/en/newsroom/press-releases/archive-2011/KLM_launches_commercial_flights_Amsterdam.html, accessed August 2011.
72. Alternative Jet Fuels, Chevron Corporation, http://www.cgabusinessdesk.com/document/5719_Aviation_Addendum._webpdf.pdf, accessed August 2011.
73. Scientific American news, "Fatted Eagle" Joins "Green Hornet" in U.S. Military's Alternative Fuels Fighter Fleet, http://www.scientificamerican.com/article.cfm?id ¼ fatted-eagle-biofuel-flightair-force-fighter, accessed August 2011.
74. D. B. Ghonasgi, E. L. Sughrue, J. Yao and X. Xu, US pat., 7, 626,063, 2009.
75. L. Li, E. Coppola, J. Rine, J. L. Miller and D. Walker, Energy Fuels,2010, 24, 1305.
76. J. C. Serrano-Ruiz, R. Luque and A. Sepulveda-Escibano, Chem. Soc. Rev., 2011, 40, 5266.
77. M. J. Haas, A. J. McAloon, W. C. Yee and T. A. Foglia, Bioresour. Technol., 2006, 97, 671.
78. G. W. Huber and A. Corma, Angew. Chem., Int. Ed., 2007, 46, 7184.
79. K. Sunde, A. Brekke and B. Solberg, Energies, 2011, 4, 845.
80. http://www.nesteoil.com/default.asp?path ¼ 1;41;540;1259;1260;16746;17731, accessed August 2011.
81. http://www.nesteoil.com/default.asp?path ¼ 1;41;540;1259;1260;13292;15693, accessed August 2011.
82. N. Ladygina, E. G. Dedyukhina and M. B. Vainshtein, Process Biochem., 2006, 41, 1001.
83. L. P. Wackett, Microb. Biotechnol., 2008, 1, 211.
84. C. Sheridan, Making green, Nat. Biotechnol., 2009, 27, 1074.
85. M. A. Rude and A. Schirmer, Curr. Opin. Microbiol., 2009, 12, 274.
86. M. Zhang, C. Eddy, K. Deanda, M. Finkelstein and S. Picataggio, Science, 1995, 267, 240.
87. G. Stephanopoulos, Science, 2007, 315, 801.
88. P. Durre, Biotechnol. J., 2007, 2, 1525.
89. S. Atsumi, T. Hanai and J. C. Liao, Nature, 2008, 451, 86.
90. K. S. Zhang, M. R. Sawaya, D. S. Eisemberg and J. C. Liao, Proc. Natl. Acad. Sci. U. S. A., 2009, 105, 20653.
91. L. A. Hazelwood, J. M. Daran, A. J. A. van Maris, J. T. Pronk and J. R. Dickinson, Appl. Environ. Microbiol., 2008, 74, 2259.

92. http://www.gevo.com, accessed August 2011.
93. D. K. Ro, et al., Nature, 2006, 440, 940.
94. P. P. Peralta-Yahya and J. D. Keasling, Biotechnol. J., 2010, 5, 147.
95. J. L. Fortman, et al., Trends Biotechnol., 2008, 26, 375.
96. N. S. Reninger and D. J. McPhee, US pat., 2008/0098645 A1, 2008.
97. http://www.amyris.com/en/markets/fuels/renewable-diesel-fuel, accessed August 2011.
98. http://bioage.typepad.com/.a/6a00d8341c4fbe53ef0133f5975d49970bpopup, accessed August 2011.
99. http://www.businesswire.com/portal/site/home/permalink/?ndmViewId¼news_viewew&;newsLang¼en&newsId¼20101101005868&div¼-2012183699, accessed August 2011.
100. http://www.reuters.com/article/2011/08/10/us-brazil-amyris-idUS-TRE7796P520110810, accessed August 2011.
101. N. S. Renninger, J. A. Ryder and K. J. Fischer, US pat., 79422940B2, 2011.
102. S. T. Withers, S. S. Gottlieb, B. Lieu, J. D. Newman and J. D. Keasling, Appl. Environ. Microbiol., 2007, 73, 6277.
103. R. Kalscheuer, T. St€olting and A. Steinb€uchel, Microbiology, 2006, 152, 3688.
104. J. D. Keasling, et al., US pat., 2010/0242345 A1, 2010.
105. E. J. Steen, Y. Kang, G. Bokinsky, Z. Hu, A. Schirmer, A. McClure, S. B. Del Cardayre and J. D. Keasling, Nature, 2010, 463, 559.
106. A. Schirmer, M. A. Rude, X. Li, E. Popova and S. B. del Cardayre, Science, 2010, 329, 559.
107. M. A. Rude, T. S. Baron, S. Brubaker, M. Alibhai, S. B. del Cardayre and A. Schirmer, Appl. Environ. Microbiol., 2011, 77, 1718.
108. http://www.ls9.com/index.html, accessed August 2011.
109. http://www.ls9.com/technology, accessed August 2011.
110. http://www.greencarcongress.com/2011/07/ls9-20110708.html#more, accessed August 2011.
111. D. E. Robertson, S. A. Jacobson, F. Morgan, D. Berry, G. M. Church and N. B. Afeyan, Photosynth. Res., 2011, 107, 269.
112. http://www.jouleunlimited.com/video/story.html, accessed August 2011.
113. M. J. Dunlop, et al., Mol. Syst. Biol., 2011, 7, 487.
114. D. Bianchi, in Biomass Catalytic Conversion to Diesel Fuel: Industrial Experience, Next Generation Biofuels, Bologna, 18 September 2009, www.ics.trieste.it/media/139853/df6504.pdf, accessed August 2011.
115. The 2 MillionMileHaul August 2008Update, Donald A. Heck, Iowa Central Community College, http://www.iowacentral.edu/mathscience/science/programs/biotechnology/research_collaboration.asp, accessed August 2011.

CHAPTER 7

Synthetic Routes to Methylerythritol Phosphate Pathway Intermediates and Downstream Isoprenoids

SARAH K. JARCHOW-CHOY, ANDREW T. KOPPISCH, AND DAVID T. FOX

7.1 BACKGROUND

With greater than 55,000 representative compounds, isoprenoids (also referred to as terpenes) are the largest and most chemically diverse family [1]. Isoprenoids are classified into groups according to the number of carbon atoms they contain, and include: hemiterpenes (C_5), monoterpenes (C_{10}), sesquiterpenes (C_{15}), diterpenes (C_{20}), triterpenes (C_{30}), tetraterpenes (C_{40}) and so forth. They play numerous functional roles in primary metabolic processes that are necessary for cell survival including cell wall and membrane biosynthesis (phytosterols), electron transport (plastoquinones), photosynthetic pigments (carotenoids), and hormones in plants (gibberellins) [2, 3]. In addition, microbial production of isoprenoids has

Synthetic Routes to Methylerythritol Phosphate Pathway Intermediates and Downstream Isoprenoids.
© 2014 Bentham Science Publishers; Current Organic Chemistry, 2014, Vol. 18, No. 8. Creative Commons Attribution Non-Commercial License (http://creativecommons.org/licenses/by-nc/3.0/)

found their way into industry as a source for nutraceuticals, pharmaceuticals and, most recently, in biofuel applications [4]. Some examples include fragrances and essential oils (menthol, eucalyptol, limonene), cosmetics (squalene), disinfectants (pinene and camphor), antimicrobials (taxol and artimesinen), food supplements (Vitamins A, E and K), food colorants (xanthanins), fuels (farnesene and botryococcenes) and dozens more of related industrial applications.

The chemistry of isoprenoid biosynthesis allows for a tremendous structural diversity within its members, both in terms of the carbon length of the molecule (e.g. monoterpenes, diterpenes) and in the connectivity observed in the isoprenoid chain after biosynthetic steps which process diphosphates (e.g. cyclization of a triterpene, condensation of two sesquiterpene diphosphates). Understandably, this chemical diversity has enabled members of the isoprenoid family to play a wide range of important biological roles throughout nature as previously mentioned. However, the biology of isoprenoid molecules is beyond the scope of this review. Generally speaking, research into isoprenoid chemistry includes work that aims to understand their role in endogenous biological pathways of relevance to human health (two such examples are the enzymology of cholesterol biosynthesis and the role of protein farnesylation in cancer progression), as well as work which is designed to isolate and characterize isoprenoid secondary metabolites of value to society (e.g. anticancer agents such as Taxol, or potential renewable fuels such as isoprene, farnesene, or botryococcenes), human health, plant and fungal biochemistry. Thus, the development of chemical routes to isoprenoid compounds aids the synthesis of some pharmacologically-valuable compounds and intermediates (many of which themselves represent challenging synthetic targets) as well as enables enzymological characterization and inhibition studies. In this review, we will focus primarily on synthetic routes that have been developed for isoprenoid intermediates for the purpose of enzyme discovery and characterization.

The chemical diversity of all isoprenoids is derived from one of two simple C_5 building blocks: isopentenyl diphosphate (IPP) and its isomer dimethylallyl diphosphate (DMAPP). These universal precursors are produced by either of two routes: the mevalonate or the 2C-methyl-D-erythritol 4-phosphate (MEP) (also known as the non-mevalonate or

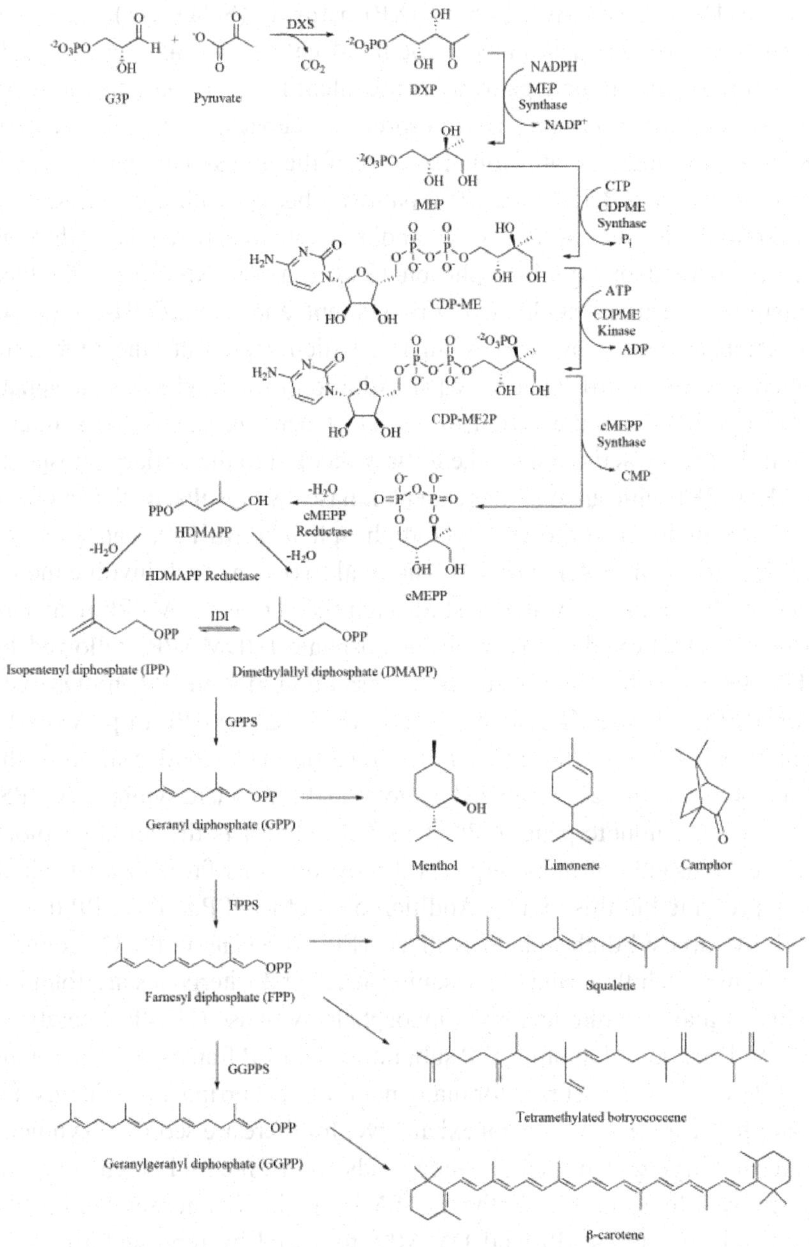

Scheme 1. The MEP pathway to isoprenoids.

1-deoxy-D-xylulose 5-phosphate, DXP) pathway (Scheme 1). Both bio-synthetic routes are distributed throughout nature, and as a general rule of thumb the mevalonate pathway is prevalent in eukaryotes and archaea, whereas the MEP pathway is widespread in eubacteria. For this review, the synthetic intermediates found as part of the mevalonate pathway will not be discussed. Briefly, the MEP pathway begins with condensation of glyceraldehyde-3-phosphate (G3P) and pyruvate to generate DXP through the action of deoxyxylulose phosphate synthase (DXS) [5]. DXP then undergoes a carbon skeleton rearrangement and a NADPH-dependent reduction catalyzed by MEP synthase, which represents the first committed step in the MEP pathway [6]. Addition of cytidyl monophosphate (CMP) to MEP via the action of a CTP-dependent (CDPME) synthase [7], followed by activation of the tertiary alcohol to the tertiary phosphate (CDPME2P) with an ATP-dependent kinase [8] results in the cyclized product, methylerythritol-2,4-cyclodiphosphate (cMEPP), catalyzed by the similarly named synthase [9]. The final two steps each involve the reductive elimination of water, first through the action of cMEPP reductase to provide hydroxydimethylallyl disphosphate (HDMAPP) followed by HDMAPP reductase that result in the formation of both IPP and DMAPP [10]. IPP isomerase (IDI) interconverts IPP and DMAPP to provide the appropriate ratios for normal cellular function [11]. Condensation of the two C_5-building blocks catalyzed by geranyl diphosphate synthase (GPPS) provides the monoterpene, GPP. This C_{10} subunit is the building block for the thousands of monoterpenes discovered thus far, a few of which are represented in this review. Addition of another IPP to the GPP framework by farnesyl diphosphate synthase (FPPS) results in the C_{15} foundation from which thousands of sesquiterpenes and triterpenes are found in Nature. Finally, geranylgeranyl diphosphate synthase (GGPPS) catalyzes the condensation of another IPP subunit to the FPP framework to provide GGPP, which is a precursor for many naturally occurring antioxidants, for example β-carotene. While not exhaustive, for there are scores of synthetic routes to varying chain-length isoprenoids and isoprenoid precursors, this review will focus on the syntheses of MEP pathway intermediates to the universal precursors, IPP and DMAPP, followed by representative synthetic routes to longer branched, unsaturated hydrocarbons that may be of interest to the reader.

7.2 ISOPRENOID PRECURSORS-METHYLERYTHRITOL PHOSPHATE (MEP) PATHWAY

Since the MEP pathway discovery to isoprenoids in the early 90s, considerable effort toward elucidating the enzymes responsible for their biosynthesis ensued [6, 12, 13]. Early studies revealed the MEP pathway was distributed in most bacteria, the chloroplasts of photosynthetic organisms, unicellular eukaryotes (e.g. green algae) and the malaria parasite [14]. Most importantly, the MEP pathway enzymes are excellent targets for the development of antimicrobial agents for there are no known human orthologs [15]. The first enzyme in the MEP pathway to isoprenoids, deoxyxylulose phosphate synthase, converts pyruvate and glyceraldehyde-3-phosphate to (–)-1-deoxy-D-xylulose 5-phosphate (DXP). Conversion of DXP to MEP through MEP synthase represents the first committed step to all isoprenoids where the mevalonate pathway is not present. Therefore, a fundamental understanding of the biological role of MEP synthase was exhaustively pursued, largely through the use of synthetic (isotopically-labeled) DX(P) substrates and corresponding substrate analogs.

Many examples for the enzyme-assisted synthesis of both DX and DXP and the corresponding isotopomers are reported in the literature [16-19]. However, preparation of large quantities of these deoxyxyluloses is not the preferred route largely due to purification problems, the requirement for expensive labeled precursors and scalability. In addition, biosynthetic production of DX(P) analogs is largely unattainable due to the stringent substrate specificity of the corresponding enzymes responsible for their biosynthesis. Therefore, synthetic preparations to these compounds are preferred due to the versatility of the chemical transformations from relatively cheap starting material, high-yielding reactions, and simple purifications.

7.2.1 SYNTHESES OF 1-DEOXY-D-XYLULOSE 4 (DX)

The first examples for the synthesis of the D-enantiomer of DX were reported by Estramareix et al. in the early 80s. Following protection and

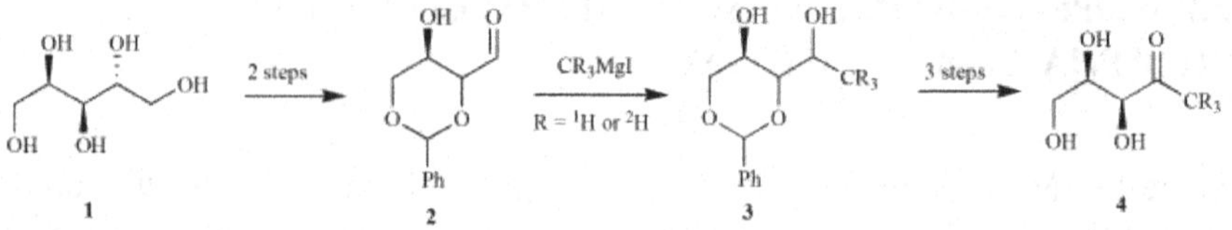

Scheme 2. Estramareix synthetic route to DX.

oxidative cleavage of D-arabinitol 1, condensation of isopropylidene-protected D-glyceraldehye and the acetaldehyde dithioacetal anion provided
a mixture of the protected threose and erythro-isomers. Separation of the
threose enantiomer from the erythro-form followed by deprotection provided D-deoxylulose 4 over three steps in 15% yield [20]. The same group
improved upon the synthesis to deoxyxylulose and the 1-^{2}H$_3$-isotopically-
labeled derivativethrough condensation of aldehyde 2 with methylmagnesium iodide to provide the 2,4-O-benzylidene-D-threose intermediate 3.
Following oxidation and deprotection, the free sugar 4 was obtained by
brominolysis of the distannylidene derivative (Scheme 2) [21, 22].

Importantly, the Grignard reaction for introduction of the methyl group
is now the most widely used method for incorporation of isotopes into
the C-1 position of deoxyxylulose (phosphate). Further, Estramareix et al.
provided the foundation for future synthetic efforts through exploitation
of the D-threose scaffold in order to set the correct stereochemistry in the
final product. As such, multiple independent routes to the free sugar using either D-threose or its oxidized derivative, D-tartrate, were employed
and overall yields were much improved [23-25], with the highest reported
yield of 68% over five steps from dimethyl-D-tartrate [26]. A generalized
reaction scheme for DX 4 syntheses from multiple laboratories starting
from either the D-tartrate 5 or D-threitol 6 derivatives are outlined below
(Scheme 3).

The previously outlined synthetic routes to DX provide a convenient
platform for incorporation of a stable isotope (^{2}H or ^{13}C) into the C-1 position of the free sugar via an aldehyde intermediate. However, incorporation of an isotopic tag at other positions along the polyol backbone is

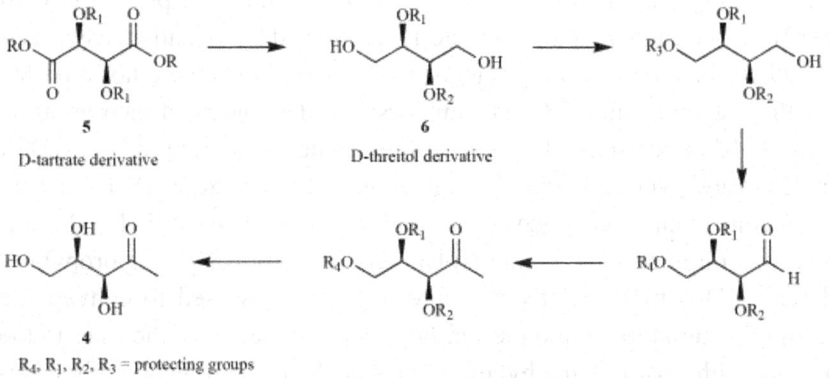

Scheme 3. Generalized synthetic routes to enantiomerically pure DX.

not feasible using these approaches. An alternative strategy was adopted by Giner et al. that enabled insertion of deuterium at the C-3 and/or C-4 positions of DX. Briefly, deuterium-hydrogen exchange of the phosphonium ylide 7 with a 10:1:1 mixture of $CH_2Cl_2:D_2O:CD_3OD$ followed by a Wittig reaction with the benzyl-protected glycoaldehyde 8 provided the conjugated 5-carbon enone 9 with 85% deuterium incorporation at the C-3 position. Subsequent Sharpless asymmetric dihydroxylation and debenzylation via hydrogenolysis provided the enantiomerically pure DX 4 in good yields [27-29]. Conversely, the deuterium was conveniently inserted

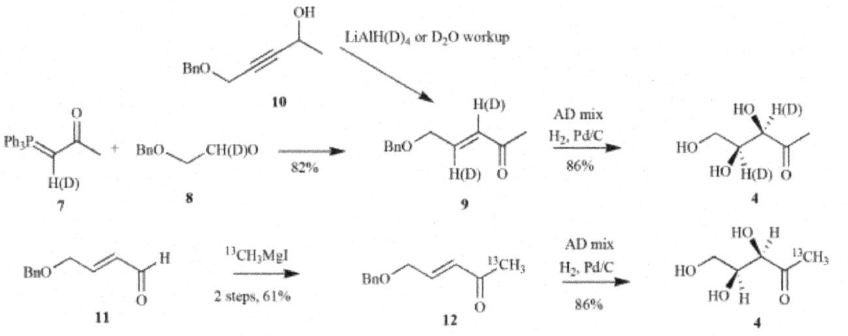

Scheme 4. Giner's syntheses of C-1 isotopically labeled DX.

into the C-4 position (~70% deuterium) by reduction of the propargyl alcohol 10 with lithium aluminum deuteride, followed by oxidation, asymmetric dihydroxylation then deprotection, although yields were not reported for the reduction step [24]. The authors also demonstrated incorporation of a ^{13}C-label into the C-1 position of aldehyde 11 to provide 1-^{13}C-DX 4 in 52% yield over four steps through enone intermediate 12 (Scheme 4).

A convenient chemoenzymatic synthesis for 1-deoxy-5,5-^{2}H$_2$-D-xylulose was outlined by Piel and Boland starting from 2,3-O-isopropylide-dimethyl-D-tartrate. Briefly, pig liver esterase was used to convert the starting material to the monoester. Following reduction of the ester to the alcohol with LiBEt$_3$D, methylation of the acid and deprotection of the isopropylidene moiety, the C-5 dideuterated DX was synthesized over four steps in 37% yield [30]. However, scalability was an issue since the first step required the use of the esterase.

7.2.2 SYNTHESES OF 1-DEOXY-D-XYLULOSE 5-PHOSPHATE 16 (DXP), ISOTOPOMERS, AND FLUORINATED ANALOGS

The first report for the synthesis of enantiomerically pure DXP was outlined by Begley et al. in 1998 [31]. The key step in this synthesis was desymmetrization of the C$_2$-symmetric dibenzylated-D-threitol by mono-

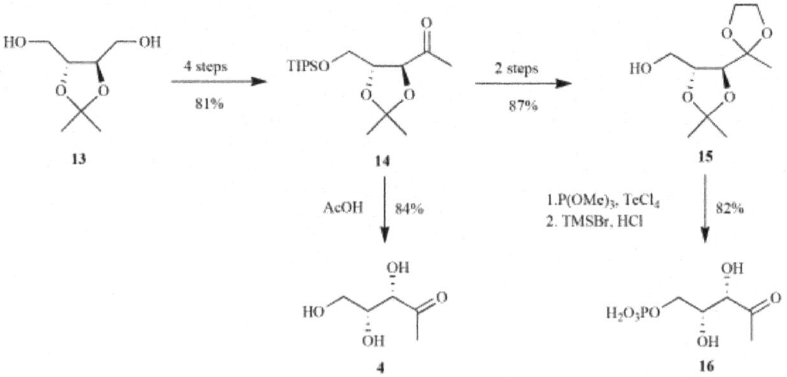

Scheme 5. Blagg and Poulter's synthetic routes to both DX and DXP from D-threitol.

phosphorylation with dibenzylchlorophosorochloridate in 40% yield. The primary alcohol was converted to the methyl ketone using modified procedures previously outlined then debenzylation via hydrogenolysis provided DXP over seven steps in 5% yield from diethyl-D-tartrate. Shortly thereafter, Blagg and Poulter reported a more efficient chemical synthesis for DXP by exploiting (–)-2,3-O-isopropylidene-D-threitol 13 as a more suitable starting material for the chemical transformations (Scheme 5). The triisopropylsilyloxy-protected DX 14 proved to be a versatile intermediate in the synthesis. Simple deprotection under acidic conditions provided DX 4 over five steps in 69% yield. Alternatively, following ketolization of the ketone with ethylene glycol and removal of the triisopropylsilyloxy-protecting group, the primary alcohol 15 was phosphorylated with trimethylated phosphite in the presence tellurium (IV) chloride. Phosphate deprotection in the presence of TMSBr followed by acidic workup and purification by cellulose chromatography provided pure DXP 16 in four steps from the fully protected DX in 71% yield [26].

MEP synthase is generally recognized as an excellent target for the development of novel antimicrobial agents. However, the enzyme-catalyzed reaction mechanism for conversion of DXP to MEP remained elusive. In order to address this gap, multiple isotopically-labeled DXP and DXP analogs were synthesized to determine the stereochemical course for the transformation. In 2003, Cox et al. reported a synthetic route to DXP from propargyl diol 17 where deuterium was incorporated into either the C3-18 or C4-19 positions, respectively, using a modified protocol reported by Giner et al. [32]. Silyl deprotection and selective phosphorylation of

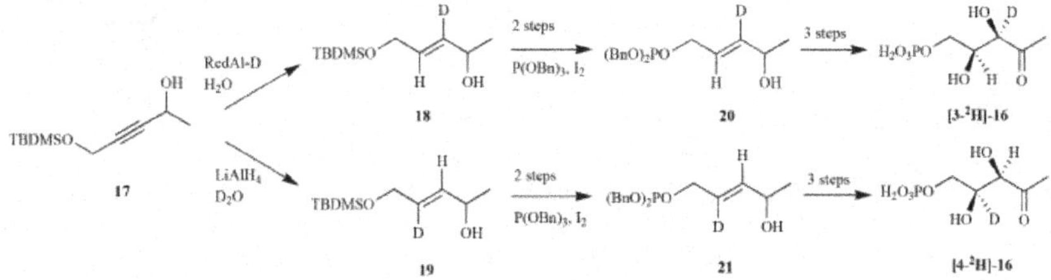

Scheme 6. Cox's synthetic routes to DXP from the propargyl alcohol.

the primary alcohol over the secondary alcohol provided 20 and 21, re-spectively. However, the overall yield and enantiopurity was suboptimal (~20%, 84% e.e.) and no demonstration for isotopic label incorporation was discussed [33]. The same authors amended their synthesis for incor-poration of the deuterium label into either the [3-^{2}H]-16 or [4-^{2}H]-16 posi-tion of DXP over eight steps in 15% or 7% yield (Scheme 6), respectively, and corroborated previous reports that MEP synthase converts DXP to MEP through a retroaldol/aldol mechanism [34-36]. Shortly thereafter, Liu et al. reported two independent routes to the identical compounds with similar yields (~8%, 95% e.e.) [37].

There are countless reports in the literature for the synthesis of DXP analogs that extend beyond the scope of this review. However, one class of DXP analogs deserves discussion since they were designed for three reasons: potential as novel inhibitors, determination of the enzyme-cat-alyzed reaction mechanism, and substrate specificity. Specifically, the fluorinated analogs of either DX or DXP were chemically synthesized for this purpose. The first report of a fluorinated deoxyxylulose analog was reported by Bouvet and O'Hagan for the synthesis of both 1-fluoro and 1,1-difluoro-1-deoxy-D-xylulose [38]. In their synthesis of the mono-fluo-rinated analog 24, the authors utilized a Wittig reaction for coupling of the silyl-protected glycoaldehyde 22 with the mono-fluorinated phosphorous ylide 23 in a reaction analogous to previous reports for the [3- and [4-2H]-DX syntheses [29]. However, Sharpless asymmetric dihydroxylation was unsuccessful thus necessitating standard dihydroxylation with osmium

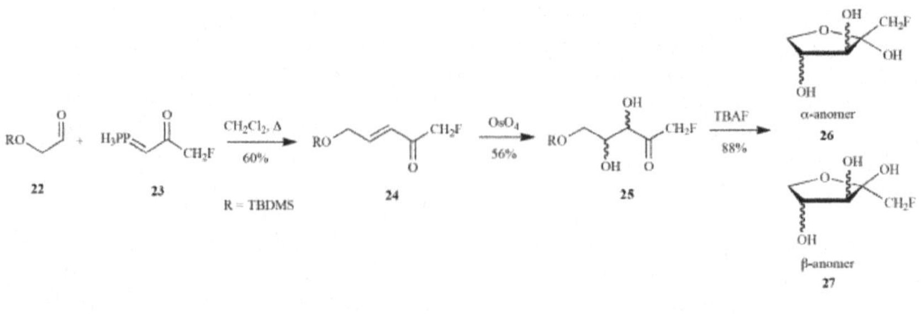

Scheme 7. O'Hagan's synthetic route to CF-DX.

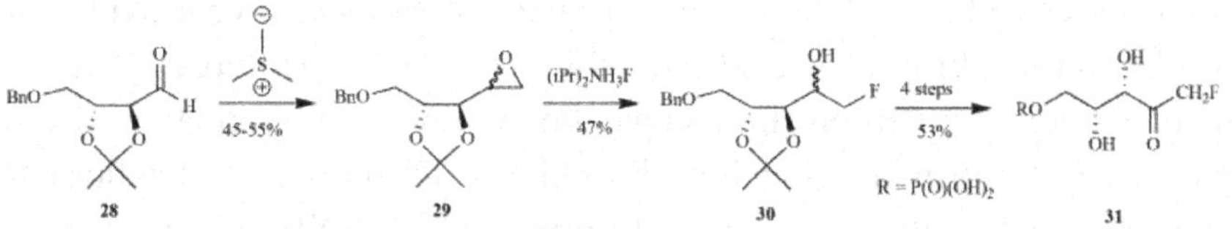

Scheme 8. Fox and Poulter's synthesis to CF-DXP from a D-threitol derivative.

tetraoxide to provide the racemic fluorinated diol 25. Deprotection of the silyl group provided the (±) C-1 monofluorinated DX in 30% yield over three steps as a 3:2 mixture of the α- and β-cyclic fluorinated anomers 26 and 27 (Scheme 7).

Using O'Hagan's syntheses as an inspiration, Fox and Poulter synthesized the 1-fluoro-, 1,1-difluoro- and 1,1,1-trifluoromethyl-1-deoxy-D-xylulose 5-phosphate analogs for mechanistic studies with MEP synthase [39]. Upon installation of the oxirane moiety into aldehyde 28, the key step in the synthesis of the 1-fluoromethyl-DXP (CF-DXP) was the regioselective ring opening of the intermediate oxirane 29 with diisopropylamine trihydrogen fluoride [40] at the C-1 position to provide the monofluorinated alcohol 30 in 45-55% yield. Debenzylation followed by regioselective phosphorylation of the primary alcohol with dibenzyl phosphoroiodidate and subsequent Dess-Martin oxidation provided the fully protected CF-DXP. Debenzylation of the phosphate by hydrogenolysis in tert-butanol instead of methanol or ethanol was necessary to avoid formation of the stable hemiketal due to the presence of the adjacent fluorine. The free acid effectively removed the isopropylidene moiety to provide CF-DXP 31 in 10% yield, with greater than 95% purity, over nine steps as a mixture of the ketone and hydrate forms (77:23) (Scheme 8). A much lengthier synthesis of the identical compound was independently reported by Liu and coworkers. In their synthesis of 1-fluoromethyl-DXP, fluorination of the primary alcohol, which was derived from D-arabinose in three steps, with triflate and tetrabutylammonium fluoride (TBAF) provided the monofluorinated sugar. Deprotection of the 1,2-diol, benzylation and reductive ring opening with sodium borohydride provided the open chain-polyol.

Regioselective benzyolation of the primary alcohol and Swern oxidation resulted in the fully protected monofluorinated deoxyxylulose. A series of protection and deprotection steps allowed for the introduction of the phosphate group with trimethyl phosphite in the presence of tellurium chloride and 2,6-lutidene to afford the protected CF-DXP. Deprotection of the phosphate methyl groups with TMSBr followed by debenzylation by hydrogenolysis provided the free acid of CF-DXP 31. This compound was estimated to be 80% pure as judged by ^{19}F NMR spectroscopy and was obtained in less than 3% yield over 14 steps [36]. In the same manuscript, Liu et al. described the synthesis of either the C-3 or C-4 monofluorinated DXP. These synthetic routes were a modified procedure as outlined for the C-1 monofluorinated DXP, required over 15 steps and the final material was obtained in low yields.

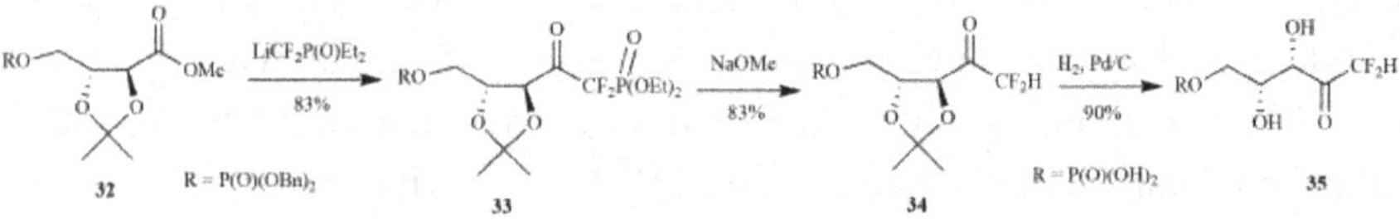

Scheme 9. Fox and Poulter's synthesis of CF2-DXP from the D-tartrate derivative.

Simple modification of O'Hagan's synthesis for the 1,1-difluoromethyl deoxyxylulose, via the identical difluoromethylenephosphonate interme-diate 33, provided 1,1-difluoro-1-deoxy-D-xylulose 5-phosphate 35 over five steps from the isopropylidene-protected D-tartrate 5. The key differ-ence was installation of the dibenzyl phosphate into the tartrate monoester 32 prior to reaction with the lithiated diethyl(difluoromethyl)phosphonate. C-P bond cleavage of the phosphonate sugar 33, after azeotropic removal of water, with sodium methoxide provided the fully protected CF_2-DXP 34 in 83% yield. Deprotection provided the difluorinated sugar phosphate 35 in 53% yield over five steps from the tartrate monoester as a mixture of the ketone/hydrate (2:98) as judged by ^{19}F NMR spectroscopy (Scheme 9).

    The key step in the synthesis of 1,1,1-trifluoromethyl-1-deoxy-D-xy-lulose 5-phosphate (CF₃-DXP) was insertion of the trifluormethyl moiety into the identical TIPS-protected aldehyde intermediate found in Blagg's synthesis of DXP [26] using (trifluoromethyl)trimethylsilane and a catalytic amount of tert-butoxide, following a procedure outlined by Olah et al. [41], to provide the bissilyl ether as a 3:1 mixture of diastereomers. Following removal of the silyl ethers with two equivalents of TBAF, regioselective phosphorylation of the primary alcohol was carried out analogous to the synthesis for the monofluorinated DXP derivative. Deprotection using the same strategy as for the mono- and difluorinated DXP analogs afforded the trifluorinated sugar phosphate exclusively as the ketone hydrate in 46% yield over eight steps and greater than 95% purity as judged by ^{19}F NMR spectroscopy.

7.2.3 SYNTHESES OF 2C-METHYL-D-ERYTHRITOL 40 (ME)

The five carbon phosphosugar 2C-methyl-D-erythritol 4-phosphate is the first committed intermediate in the mevalonate-independent route, or MEP pathway, to isoprenoid compounds. The methylerythritol sugar is also known to be recognized, imported and incorporated into (after in vivo phosphorylation to form MEP) the isoprenoid products of various prokaryotes [42, 43]. Thus, synthetic preparations of ME and isotopically-enriched ME proved crucial to the efforts to identify the enzymes responsible for IPP and DMAPP biosynthesis through the MEP pathway [44, 45]. More recently, preparations of ME were instrumental in the discovery of its ability to act as a small-molecule stimulant of isoprenoid biosynthesis [46], and in studies of its influence upon plant development [47]. The identification of MEP synthase (which makes MEP from DXP) and

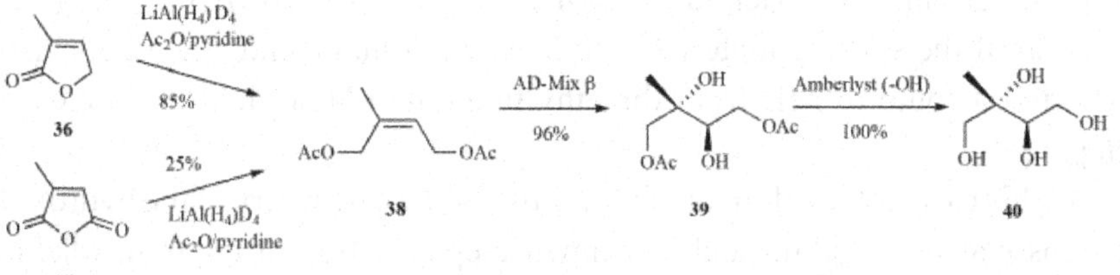

Scheme 10. Rohmer's synthesis of ME from 3-methylfuran-2(5H)-one or citraconic anhydride.

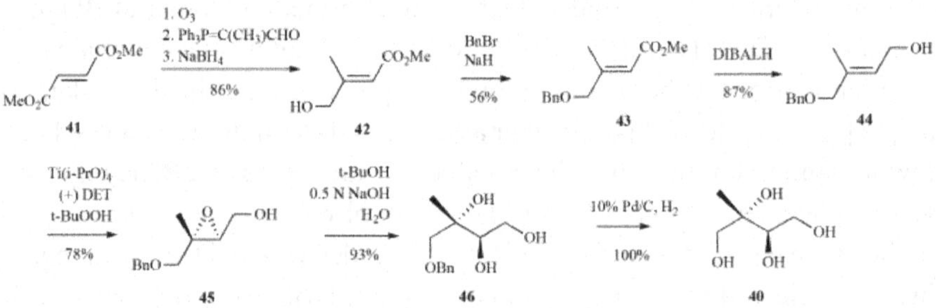

Scheme 11. Fontana's synthesis of ME from dimethylfumarate.

downstream enzymes provided an impetus for synthetic preparations of MEP itself [48].

Although ME is produced in the leaves and flowers of some plants [49], and abiotically through the atmospheric oxidation of isoprene [50-52], discovery efforts in the MEP pathway necessitated access to multi-milligram quantities of the sugar. Prior to these efforts, no dedicated chemical syntheses of the sugar existed, although some syntheses of derivatives of ME had been reported [53]. A facile route to ME was reported by Duvold et al. in 1997, wherein ME was produced in 85% yield (80% e.e.) from 3-methylfuran-2(5H)-one 36 or 25% from citraconic anhydride 37 in four steps [54].

Sharpless asymmetric dihydroxylation was used to introduce the C-2 and C-3 hydroxyl groups with the correct stereochemistry into cis-olefin 38 to provide the diacetate 39 protected ME in excellent yield. Deprotection provided ME 40 in 80% yield over three steps from the furan 36 (Scheme 10) [55]. This route also provided for the incorporation of heavy atom labels into the backbone of ME (as [1,1,4,4-^{2}H]-ME) through reduction of the starting material with LiAlD$_4$. Fontana and coworkers also reported a route to ME from dimethylfumarate (31% yield, 84% e.e., 6 steps) [56].

In this case, establishment of the correct stereochemistry of the hydroxyl groups was afforded through asymmetric epoxidation of the trans-alkene 44 to give epoxide 45 followed by ring opening with sodium hydroxide to provide the benzyl-protected ME 46. Reduction of 46 via hydrogenolysis

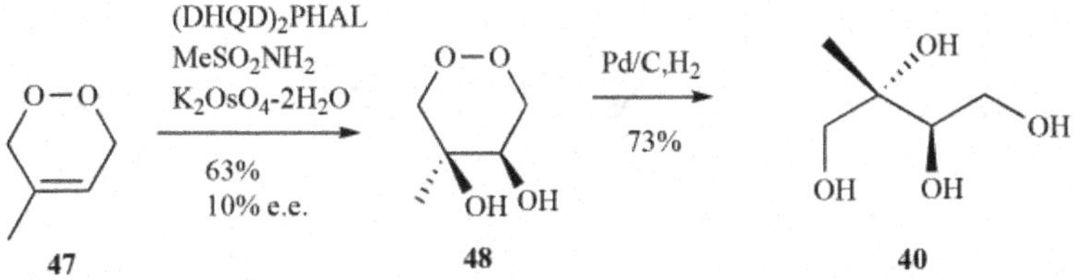

Scheme 12. Taylor's synthesis of ME from 4-methyl 1,2-dioxine.

over palladium gave ME 40 in moderate yield over eight steps (Scheme 11). Heavy atom incorporation is also possible using this route, either in the form of deuterium (reduction with LiAlD$_4$ or similar reagent) or ^{13}C (use of U-^{13}C-dimethylfumarate). Asymmetric hydroxylation or epoxidation of 3-methylbut-2-enyl-1,4-diols to introduce the stereochemistry of the C-2 and -3 hydroxyl groups proved to be a facile and useful means that also found utility in the synthesis of MEP itself and a trifluorinated-derivative of ME [57]. Even more recently, a strategy to ME and its stereoisomers was published reliant upon this means to introduce hydroxyl functionalities [58], although the scheme largely utilizes methods previously reported in the synthesis of MEP [59, 60]. In another compelling application of the asymmetric dihydroxylation reaction in the context of branched carbohydrate synthesis, Taylor and coworkers described a two-step synthesis to ME 40 from 4-methyl 1,2-dioxine 47 (Scheme 12) [61].

Central to this approach was generation of the dioxine intermediate diol 48 with the correct stereochemistry through photooxidation of a 1,3 butadiene precursor. Although the enantiomeric excess for the synthesis of ME using this method was somewhat low, the authors were able to produce a number of ME analogs with good yield and high enantioselectivity.

There are several reports for the enantiopure synthesis of ME. The use of protected carbohydrates as precursors to ME is a particularly effective means to ensure enantiopurity. Kis et al. were among the first to report such a strategy in their synthesis of ME and MEP in 2000, wherein 1,2:5,6-di-O-isopropylidene-D-mannitol was converted to ME in 10 steps or MEP in 13 steps [62]. In 2003, Hoeffler et al. reported the synthesis of

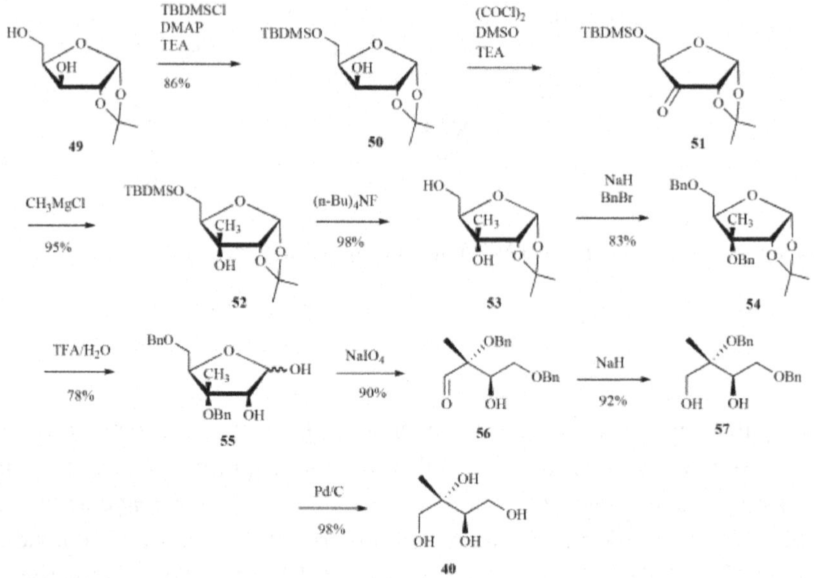

Scheme 13. Hoeffler's synthesis of ME from 1,2-O-isopropylidene-α-D-xylofuranose.

enantiopure ME in 43% yield over 8 steps from 1,2-O-isopropylidene-α-D-xylofuranose 49 [63, 64]. In the authors' strategy, the 2-methyl functionality in the final product was introduced into a TBDPS-protected 1,2-O-isopropylidene-α-D-xylofuranose 50 by oxidation of the secondary alcohol to ketone 51, followed by methyl-Grignard addition to the less hindered face of the carbohydrate ring to provide the carbon backbone 52 in the correct stereochemistry. Multiple high yielding deprotection/protection steps (53-55) followed by periodate cleavage and aldehyde 56 reduction then debenzylation of the penultimate intermediate 57 resulted in the final ME 40 product (Scheme 13).

In this manner, the authors were able to exclusively obtain the desired diastereomer of the branched chain sugar. As with similar reports, incorporation of deuterium or tritium labels into the final product was enabled in the strategy.

Coates and coworkers have reported a versatile synthesis of ME, MEP and cMEPP from D-arabitol as the starting material [65, 66]. In

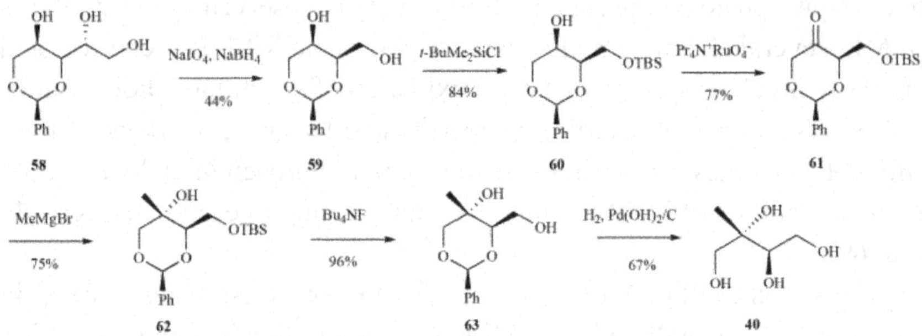

Scheme 14. Coates' synthesis of ME from D-arabitol.

the authors' strategy, benzylidene-protected D-arabitol **58** was first converted to 1,3-benzylidene-D-threitol **59** and subsequently transformed to (2S,4R)-cis-2-phenyl-4-tert-butyldimethyl-silyloxy-1,3-dioxan-5-one **61** in two steps. The dioxanone intermediate was selectively alkylated on the axial face (20:1) with methylmagnesium bromide to form the methyltetrol **62** core of ME and related metabolites. The final product is generated by silyl-deprotection to provide **62** then hydrogenolysis to enantiopure ME **40** (8% overall yield from D-arabitol in 9 steps), or alternatively **62** may be further modified to form MEP or cMEPP (Scheme 14). Furthermore, the novel methylerythritol analogue, 1-amino-1-deoxy-2C-methylerythritol was also produced using this approach [65]. Lai and coworkers have reported modifications to this dioxanone strategy which resulted in a shorter synthesis with a slightly higher yield (7 steps, 11.5% overall yield) [67]. A similar strategy of diastereocontrolled addition of an alkyl Grignard reagent into a protected ketone was used by Chattopadhyay and coworkers to provide a straightforward and economical route to ME and other methyltetrol isomers [68].

7.2.4 SYNTHESES OF 2C-METHYL-D-ERYTHRITOL 4-PHOSPHATE 71 (MEP)

Similar to other MEP pathway intermediates, MEP was previously synthesized using an enzymatic approach through condensation of glyceraldehyde-3-phosphate and pyruvate [7, 55]. As was observed in the synthesis of ME, several chemical routes to the 4-phosphate MEP are reported, and similarly involve asymmetric dihydroxylation of substituted diols or utilization of carbohydrate starting materials. Poulter and coworkers demonstrated the synthesis of MEP using the former approach in 2000 and 2002 in their synthesis of MEP as a monosodium salt and free acid, respectively [60, 69].

The sodium salt of MEP was synthesized from 1,2-propane diol 64 in 7 steps with an overall yield of 32% (78% e.e.). Following silyl-protection 65 of the primary alcohol and oxidation to provide ketone 66, the cis-configuration was installed through a Wittig-Horner olefination reaction to give cis-olefin 67 in excellent yield. Reduction of the ester and phosphorylation of primary alcohol 68 provided the phosphotriester 69 in 84% yield over the two steps. Dihydroxylation of the olefin 69 was conducted using the method of Sharpless after installation of the phosphotriester, which necessitated buffering of the standard reaction conditions with $NaHCO_3$, to provide the protected MEP 70 in moderate yield [55, 70]. Conversion of the phosphomethylesters to the corresponding acids was accomplished

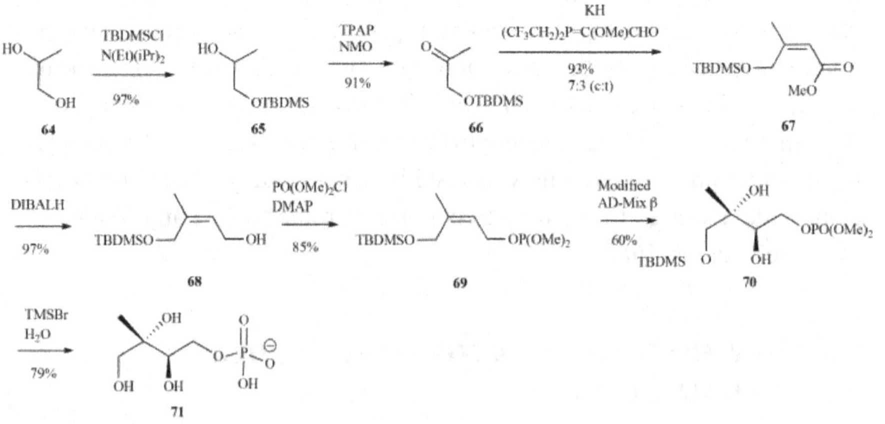

Scheme 15. Koppisch and Poulter's synthesis of MEP from 1, 2-propane diol.

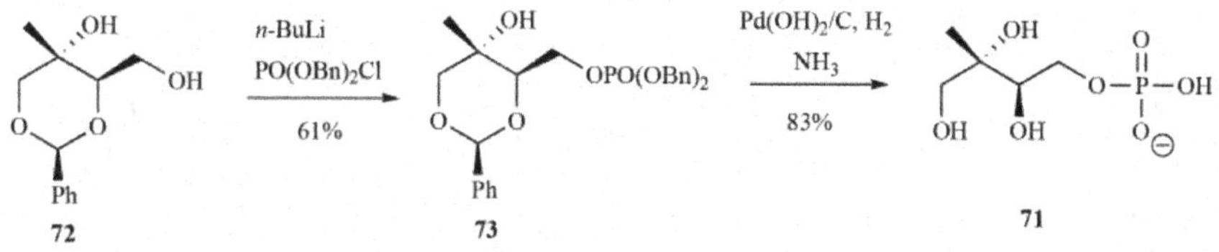

Scheme 16. Coates'synthesis of MEP from D-arabitol.

with TMS-Br under acidic conditions, and MEP 71 as the monosodium salt was formed upon neutralization (Scheme 15). Formation of MEP as the free acid is accomplished by introducing the phosphate functionality as a benzyl-protected triester, which is ultimately removed via hydrogenation over Pd/C. Hoeffler et al. used this strategy to install the phosphate moiety into MEP ultimately from an 1,2-O-isopropylidene-α-D-xylofuranose starting material using an approach inspired from the authors earlier work toward the preparation of ME (as discussed above) [64]. Similarly, both Fontana [71] and Koppisch and Poulter [69] utilized benzyl ester protected phosphate groups to form the free acid of MEP from approaches that were inspired by previous efforts of their respective teams.

Synthesis of MEP from various protected carbohydrates, as noted above, has been reported by a number of teams. Enantiopurity of the final product is ensured by utilization of the carbohydrate starting material, and their use also provides a convenient means to introduce isotopic labels. Enantiopure preparations of MEP as the sodium salt from 1,2:5,6-di-O-isopropylidene-D-mannitol proceeded in 9% yield over 14 steps (as compared to 32% over 7 steps) [72] and preparation of the free acid from 1,2-O-isopropylidene-α-D-xylofuranose proceeded in 32% yield over 7 steps (compared to 27% yield over 5 steps) [64]. More recent syntheses from carbohydrate-based starting material have also provided facile access to other metabolites based on an ME backbone (such as MEP or cMEPP) [65, 66] or in the synthesis of derivatives of MEP [73]. As mentioned earlier, a deprotected intermediate of the methylated dioxanone 72 in Coates and coworkers synthesis of ME from D-arabitol is readily phosphorylated

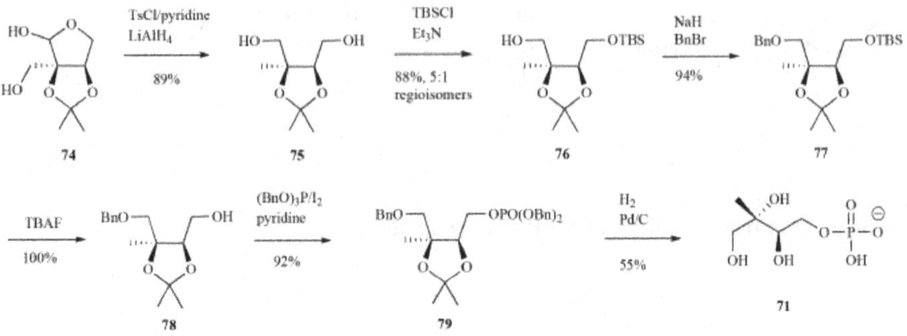

Scheme 17. Koumbis' synthesis of MEP from D-arabinose acetonide.

by lithiation with n-BuLi followed by reaction with dibenzyl chlorophosphoridate to provide the benzyl-protected MEP 73. Removal of the benzyl protecting groups via hydrogenation followed by treatment with ammonia provided MEP as the ammonium salt in good yield 71 (Scheme 16) [65].

In 2007, Koumbis and coworkers devised a synthesis of MEP from D-arabinose acetonide 74 [74]. In this case, the authors converted the starting material into an erythritol acetonide by tosylation of the free hydroxyls followed by LiAlH$_4$ reduction to form the 1, 4 diol 75. Monoderivitization of the diol with various protecting groups was performed using a tin-mediated procedure developed for 1,3- and vicinal 1,2-diols resulted in a high ratio

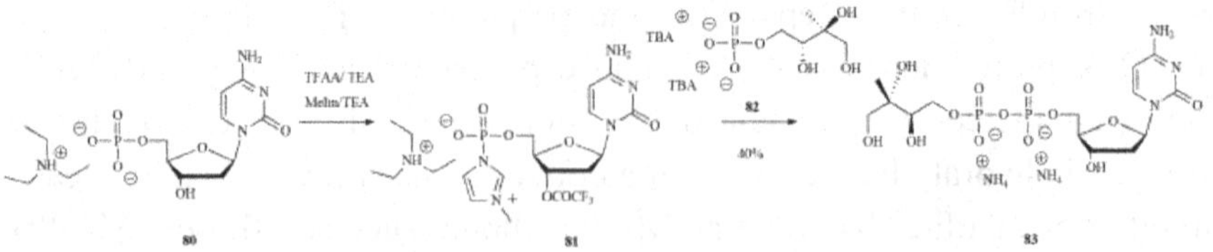

Scheme 18. Koppisch and Poulter's synthesis of CDPME from MEP and CMP.

(5:1) of the desired monoprotected alcohol 76 [75]. Following benzylation of the free alcohol to to provide the fully protected 77, silyl-deprotection 78 and phosphorylation as its benzylated derivative 79 was accomplished in over 90% yield. Deprotection via hydrogenation ultimately provided MEP 71 as its free acid in 37% yield over 7 steps (Scheme 17). Crick and coworkers used a similar approach to synthesize enantiopure MEP as its free acid from 1,2-O-isopropylidene-α-D-xylofuranose [76, 77]. Similar to the syntheses described earlier, a dibenzyl protected phosphate moiety was key to formation of MEP as its corresponding free acid.

7.2.5 SYNTHESES OF 4-DIPHOSPHOCYTIDYL-2C-METHYL-D-ERYTHRITOL 83 (CDPME)

Relatively fewer synthetic procedures to CDPME 83 exist, and with the exception of a fully enzymatic approach [7], most depend significantly upon prior routes toward the synthesis of the free acid of MEP. Koppisch and Poulter reported the first synthetic preparation of CDPME in 2002 (Scheme 18) [60]. In the authors' synthesis, the key step was coupling of MEP 82 (as a tributylammonium salt) with cytidine monophosphate 81 (as a triethylammonium salt) via activation of the phosphate 80 with trifluoro-acetic anhydride and methylimidazole, respectively. This phosphoramidite based approach was originally developed in a synthesis of UDP-galactofu-ranose, but has been incorporated into most reported syntheses of CDPME as well [78].

Upon workup and purification using size exclusion chromatography, the ammonium form of CDPME 83 was synthesized from MEP in a single pot with an overall yield of 40%. This phosphoramidite based coupling strategy has been employed in most CDPME syntheses reported to date, with the most notable exception being an enzymatic coupling strategy to form CDPME with recombinant CDPME synthase (IspD) and synthetic MEP prepared from 1,2-O-isopropylidene-α-D-xylofuranose [77]. In a similar fashion, D-arabitol was used in the synthesis of CDPME by Lai and coworkers through phosphoramidite coupling of alkylammonium salts of MEP and CMP [67].

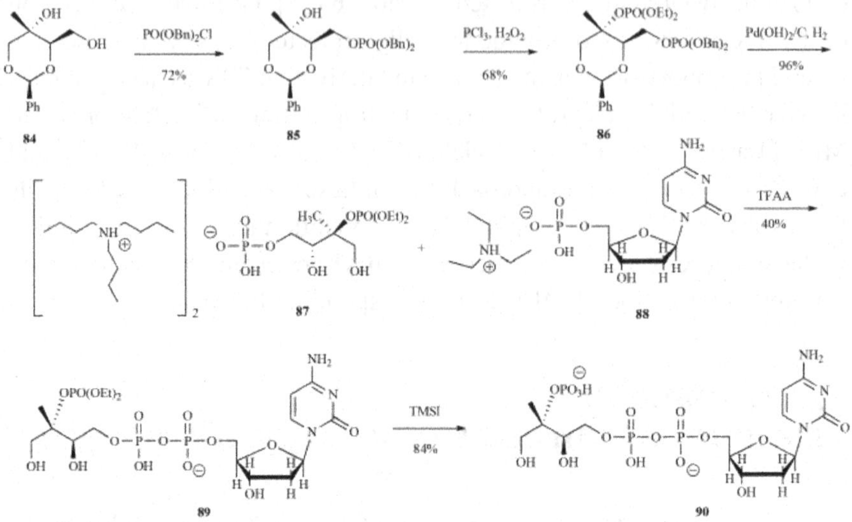

Scheme 19. Coates'synthesis of CDP-ME2P.

7.2.6 SYNTHESES OF 4-DIPHOSPHOCYTIDYL-2C-METHY-D-ERYTHRITOL 2-PHOSPHATE 90 (CDP-ME2P)

In this synthesis, the hydroxyl groups of benzylidene diol 84 (from a previous synthesis [66]) were phosphorylated, in a stepwise manner, with first dibenzyl phosphochloridate 85 then phosphorous trichloride in the presence of ethanol to provide the double phosphotriester 86 in good yield. Benzyl deprotection was performed by hydrogenolysis in one step to provide the diethylphosphate protected ME2P 87 in near quantitative yield. Triethylammonium cytidine monophosphate 88 was then activated to the phosphoramidite and coupled with the phosphoramidite of 87 to provide the phosphate-protected CDP-ME2P 89 in moderate yield. The diethyl phosphate moiety was deprotected using TMSI to provide CDP-ME2P 90 in 16% overall yield over five steps (Scheme 19) [76].

Crick and coworkers also reported a route to the MEP intermediate CDP-ME2P from D-arabitol. In this work, the authors extended the synthetic route to ME/MEP developed by Urbansky et al. to form several

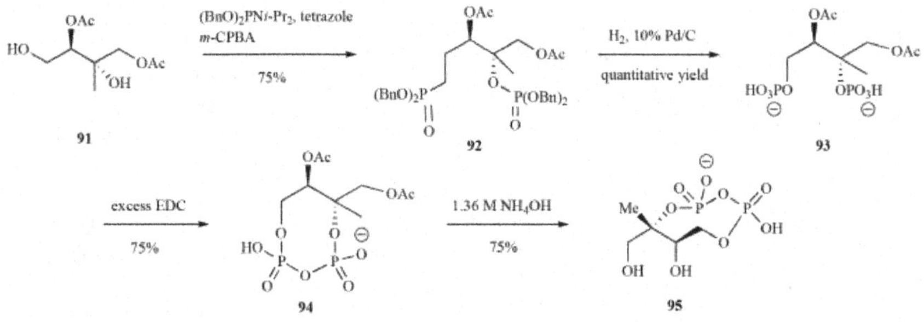

Scheme 20. Giner and Ferris synthesis of cMEPP.

benzylphosphoester or methylphosphoester derivatives of the aforementioned authors' key dioxanone intermediate. In one case, the tertiary
hydroxyl of the dioxanone intermediate was converted to a dibenzylphos-
phoester with PCl$_3$ and benzyl alcohol, while its primary hydroxyl was
transformed into a methylphosphoester with dimethyl phosphochloridate/
n-BuLi. Coupling of this product to CMP through formation of the CMP
phosphoramidite was accomplished in the manner of previous syntheses
albeit in low yield. A second strategy, wherein the bisphosphorylated dioxanone intermediate was formed as a benzylphosphoester (at the primary hydroxyl group) and ethylphosphoester (at the tertiary hydroxyl) was formed
first, and upon hydrogenation over Pd/C, 2,4-bisphosphomethylerythritol
was formed as the 2-ethylphos-phoester. Phosphoramidite coupling of this
material with CMP proceeded smoothly, and upon deprotection yielded
CDP-ME2P in good yield [76].

7.2.7 SYNTHESES OF 2C-METHYLERYTHRITOL 2,4-CYCLODISPHOPHATE 95 (CMEPP)

Cyclization of 4-diphosphocytidyl-2C-methyl-D-eryth-ritol 2-phosphate
by cMEPP synthase generates 2C-methyl-D-erythritol 2,4-cylodiphos-
phate (MECDP) [9]. An efficient synthesis of cMEPP was reported by
Giner and Ferris using 2C-methyl-D-erythritol 1,3-diacetate 91 as the
starting material. The authors reported simultaneous phosphorylation of

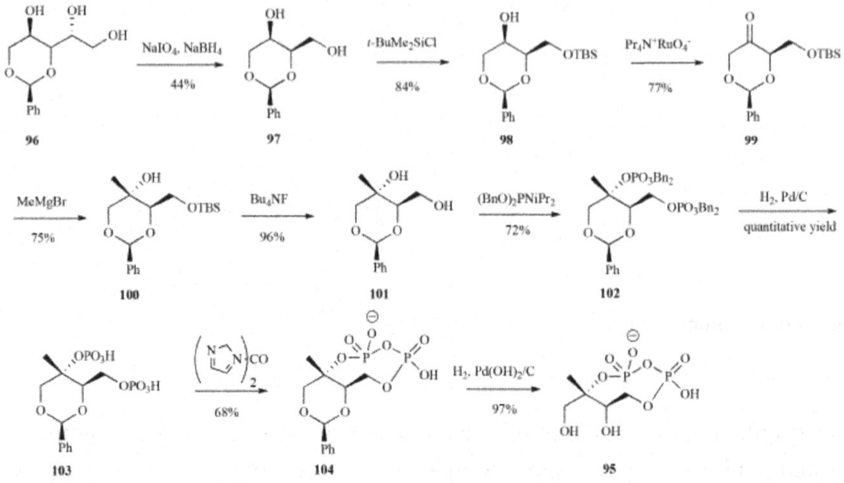

Scheme 21. Coates' synthesis of cMEPP from the protected D-threitol.

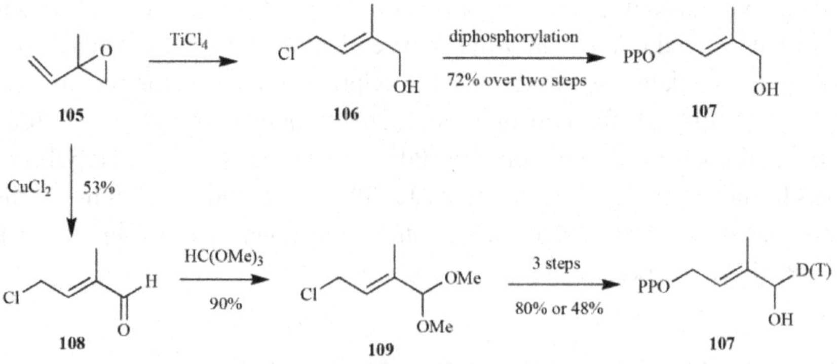

Scheme 22. Hecht's synthesis of HDMAPP and isotopomers from the vinyloxirane.

both the primary and tertiary alcohol through diisopropyl dibenzyl phosphite under basic conditions and in situ oxidation to provide the dibenzyl protected phosphate 92 in 75% yield. Hydrogenation proceeded cleanly to provide the deprotected phosphate 93 followed by phosphate coupling via

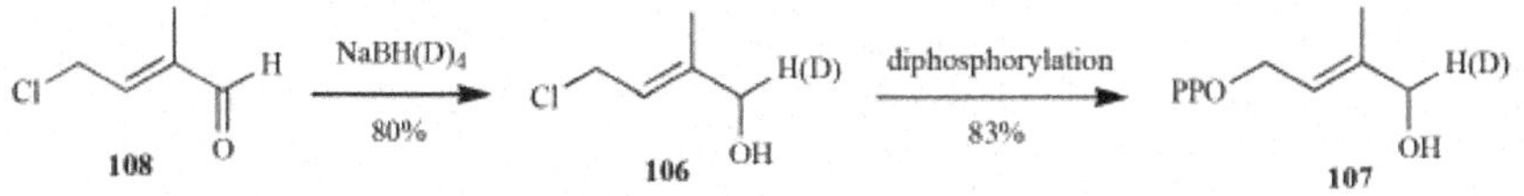

Scheme 23. Fox and Poulter's synthesis of HDMAPP via the chloroaldehyde intermediate.

carbodiimide cyclization 94 and acetate removal under careful basic conditions to provide the cyclic diphosphate 95 in four steps with 42% overall yield from the diacetate (Scheme 20) [79].

A second, lengthier synthesis began with 1,3-benzylidene-D-threitol 96. Following periodate cleavage, 97 was selectively protected with TBSCl at the primary alcohol to provide the monoether 98. A phase-transfer-promoted RuO_4 oxidation provided ketone 99 followed by reduction with methylmagnesium bromide 100 and removal of the protecting group yielding the benzylidene protected ME 101. Double phosphorylation using phosphoramidite coupling instilled the two benzyl-protected phosphate esters 102 and selective debenzylation of the phosphate esters via hydrogenation provided the bis-phosphomonoester 103, which was subsequently cyclized using 1, 1'-carbonyl diimidazole to provide the penultimate intermediate 104 [65, 66]. The final product, cMEPP 95, was generated by simple hydrogenation over palladium hydroxide (Scheme 21). This method required 9 steps as opposed to the 4 steps required in the previous synthesis and was also lower yielding (10% vs. 42%).

7.2.8 SYNTHESES OF (E)-4-HYDROXYDIMETHYLALLYL DIPHOSPHATE 107 (HDMAPP)

This straightforward synthesis was reported independently by several research groups in rapid succession upon its initial discovery in 2002 [80-86]. Of the seven, three utilized a strategy to incorporate an isotopic tag in either penultimate or final step of the synthesis. The most efficient syn-

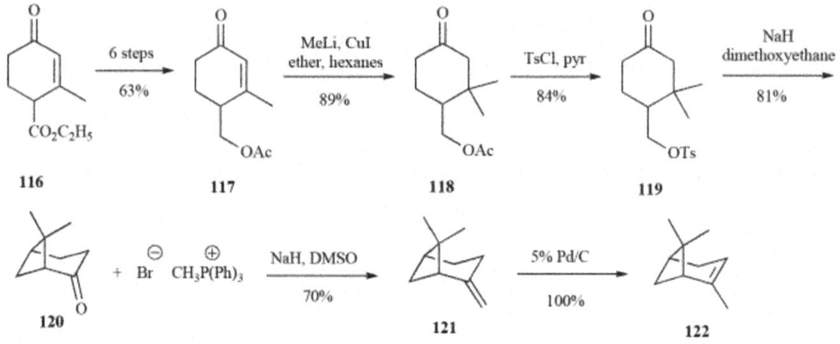

Scheme 24. Davisson's synthesis of both IPP and DMAPP.

thesis of non-radiolabeled HDMAPP was obtained in two steps from the commercially available 2-methyl-2-vinyloxirane 105 in 72% overall yield (Scheme 22). Importantly, the (E)-geometry of the chloroalcohol 106 was preferred (97:3) upon ring opening in the presence of titanium (IV) tetrachloride as determined from nuclear Overhauser effects spectroscopy (NOESY) followed by diphosphorylation to provide HDMAPP 107. From the identical starting material 105, ring opening in the presence of cupric (II) chloride provided the chloroaldehyde 108 in moderate yield. Following conversion of the aldehyde to the methylacetal 109, either a deuterium or tritium was introduced in the final step to provide labeled HDMAPP in 23% or 15% yield, respectively [85].

Fox and Poulter demonstrated the (E)-chloroaldehyde as a versatile intermediate for insertion of an isotopic label to C-1 in the penultimate step of the synthesis (Scheme 23). The overall yield for reduction of the intermediate chloroaldehyde 108 to the chloroalcohol 106 followed by phosphorylation provided HDMAPP 107, with purification of the intermediate olefin, were 66% [87].

Isopentenyl diphosphate and dimethylallyl disphosphate (IPP and DMAPP)-Both IPP and DMAPP are derived from HDMAPP, where the MEP pathway is operative, by reductive elimination of water [88] and represent the requisite building blocks to all isoprenoids. The synthesis of

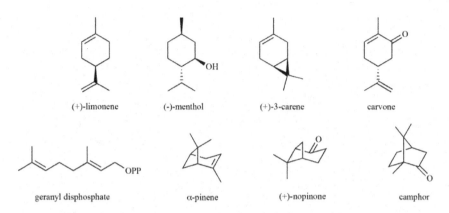

FIGURE 1: Examples of naturally occurring acyclic, monocylic and bicyclic monoterpenes.

both compounds was optimized in the mid-80s by Davisson and Poulter during their investigations for improving the diphosphorylation step of the allylic 110 or non-allylic 111 prenyl alcohols over previously published reports [89]. Briefly, the authors found that when pyrophosphoric acid was treated with tetra-n-butyl ammonium hydroxide and immediately lyophilized, tris(tetra-n-butyl) ammonium hydrogen pyrophospate was obtained on a multigram scale at near quantitative yields for the displacement reaction with the corresponding activated allylic 112 and nonallylic 113 prenyl moieties. Following cation exchange and cellulose chromatography, the ammonium form for prenyl disphosphates, 114 and 115, were typically obtained in excess of 75% yield on gram scales and could be stored for long durations under cool temperatures with minimal decomposition (Scheme 24) [90, 91].

7.3 ISOPRENOIDS

7.3.1 MONOTERPENES

Monoterpenes, which comprise the C_{10} class of isoprenoids, are largely derived from condensation of IPP to DMAPP in either a 'head-to-tail' or

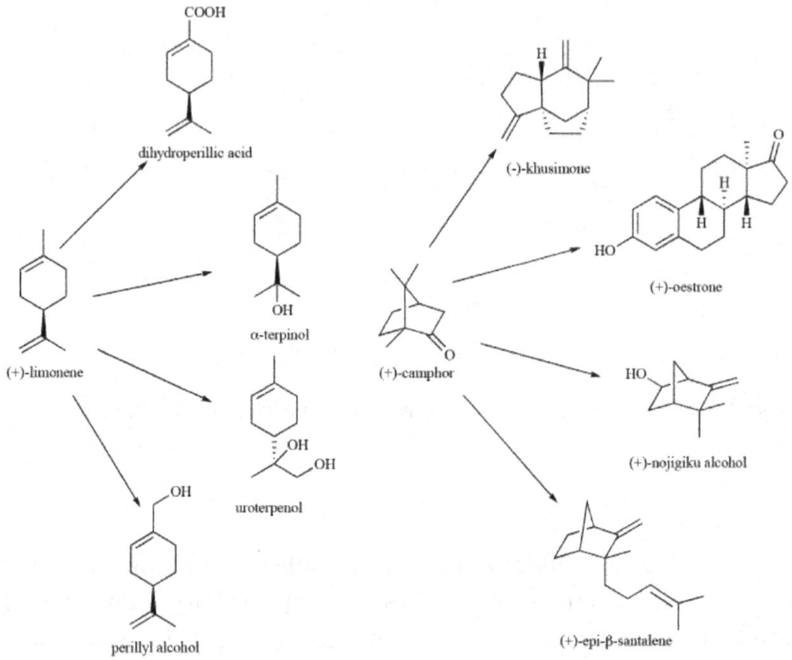

FIGURE 2: Common chemical transformations from (+)-limonene and (+)-camphor.

'non-head-to-tail' fashion or, alternatively, through condensation of two DMAPP monomers. These enzyme-catalyzed reactions ultimately result in the extremely large structural diversity observed within this class of compounds [92-94]. They are often found in pesticides, essential oils, fragrances, for use in homeopathic medicine, and are commonly used building blocks to higher-value chemicals, thus demonstrating their utility for industrial applications [95, 96]. There are countless reports in the literature for the chemical syntheses of the acyclic, mono- and bicyclic monoterpenes and the reader is directed to several sources for reviewing synthetic approaches to these industrially useful compounds [97-101]. Some common monoterpenes are found in (Fig. 1) and a few representative examples for their syntheses are presented.

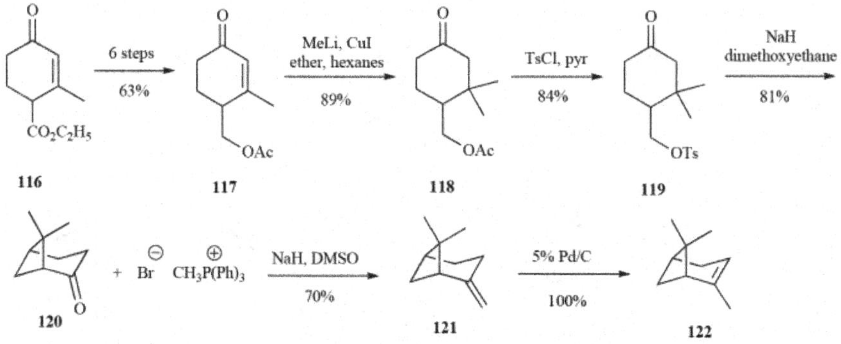

Scheme 25. Fallis' synthesis of both β-pinene (121) and α-pinene (122).

The cyclic monoterpene, (+)-limonene, is derived from cyclization of GPP and has proven to be an essential building block toward the synthesis of dozens of closely related compounds. Further, limonene is primarily accessed through cold-pressed orange oil on large scales, making this monoterpene a rather inexpensive chiral starting material suitable for chemical synthesis, and is therefore utilized as building blocks to dozens of related compounds. Representative examples of these chemical transformation are found in a review by Thomas and coworkers and some basic chemical transformations from limonene found in (Fig. 2) [102]. Another common bicyclic monoterpene, camphor, is found in the wood of camphor trees and

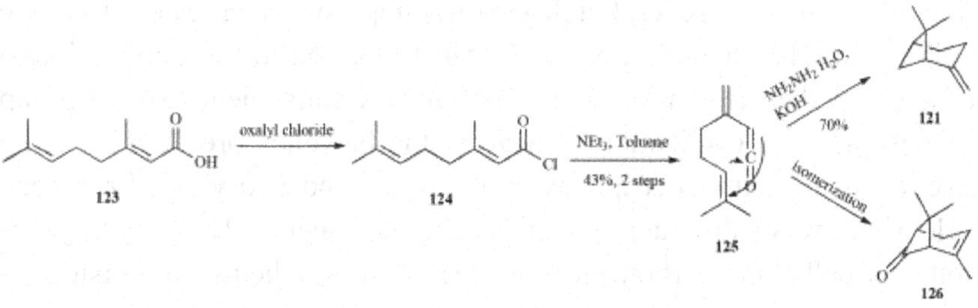

Scheme 26. Snider's synthesis of both β-pinene and chrysanthenone.

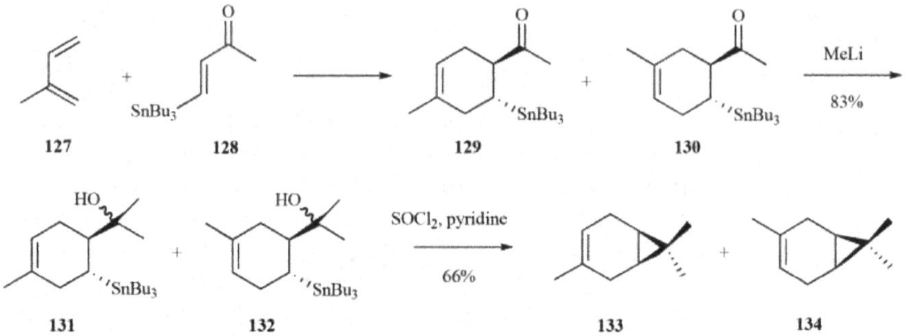

Scheme 27. Johnson's synthesis of a stereoisomeric mixture of 3-carene.

is most often utilized as a plasticizer in certain materials, for therapeutics, as a wax, and as an aromatic fragrance in essential oils. The synthesis of camphor from α-pinene was first demonstrated in 1922 and served as a building block for hundreds of chemical transformations to related iso-prenoid derivatives [103].

Among bicyclic monoterpenes, α- and β-pinenes are among the most widely distributed of the isoprenoid family and are found in the essential oils of most trees. They are currently of industrial importance as solvents and for the commercial preparation of camphor and α-terpinol [104]. However, the bicylic system of pinene presented a synthetic challenge. Fallis and coworkers addressed this through utilization of a bicyclic heptane system common to many monoterpenes within this structural class of iso-prenoids. Following several high-yielding steps to convert ester 116 to the well-known Hagemann's ester 117, transformation to the dimethyl keto acetate 118 then tosylation to 119 provided a convenient leaving group for base-assisted cyclization to provide the bicyclic core 120. A Wittig reaction with methylenetriphenylphosphorane and 120 yielded β-pinene 121, which was subsequently isomerized to α-pinene 122 by hydrogenation over palladium/carbon (Scheme 25) [105]. All chemi-cal transformations were in excess of 70% yield, making this strat-egy to this class of compounds useful for large scale production.

Another, more facile synthesis to β-pinene was developed by Snider et al. using a stereospecific [2+2] cycloaddition of ketenes to alkenes. In this synthesis, the authors converted geranic acid 123 to the corresponding acid chloride 124. This was a versatile intermediate since the presence of the electron-withdrawing halogen enabled formation of the ketene 125 with the necessary regiochemistry for subsequent cyclization to the vinyl ketones. β-Pinene 121 was generated from the ketene through a Wolff-Kishner reduction or to the monoterpene, chrysanthenone 126, through simple isomerization (Scheme 26) [106].

The bicyclic monoterpene, carene, is one of the major constituents in turpentine [107]. In one synthesis, the cyclic hexene group was constructed from isoprene 127 and ketone 128 by a Diels-Alder reaction to form the methylated cyclohexene regioisomers 129 and 130. Following methyl addition to the ketone to provide a racemic mixture of tertiary alcohols 131 and 132 the bicyclic product, 3-carene 133, and the methyl regioisomer 134 were obtained using thionyl chloride (Scheme 27) [108]. This synthesis is useful since 3-carene is used for the construction of other monoterpenes and cyclopropanes. A detailed account of synthetic approaches to new monoterpenes using carene as the building block are found in a recent review [109].

Whereas there are countless numbers of chemical syntheses to monoterpenes and their structural derivatives, many are traced back to a few, basic mono- and bicyclic compounds. There are dozens of reviews, manuscripts and books readily available that enable a detailed understanding of the evolution for chemical approaches to a myriad of monoterpenes and those presented in this review are meant to serve as a nucleation point for future synthetic efforts.

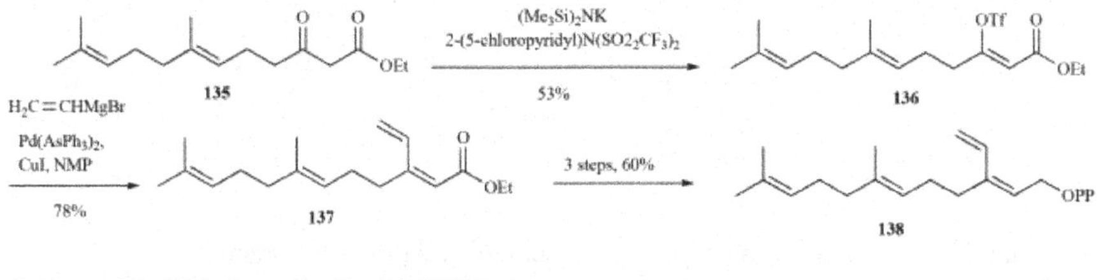

Scheme 28. Gibbs' synthesis of 3-VFPP.

7.3.2 SESQUITERPENES

Sesquiterpenes, which comprise the C_{15} class of isoprenoids, are predominantly derived from the condensation of an additional IPP monomer to the C_{10} monoterpene framework to provide farnesyl diphosphate (FPP). They are a structurally diverse group where all are derived from FPP via differing enzyme-catalyzed cyclization reactions and carbon skeleton rearrangements. They are commonly found as essential oils in plants in the mono-, bi- and tricyclic forms and have been shown to have a wide range of biological activities [110]. However, in spite of the expansive structural diversity, it is the acyclic form that has garnered the most attention from a synthetic chemistry perspective. Briefly, previous studies have demonstrated that mammalian Ras proteins, when farnesylated, result in cellular signal transduction leading to normal cell growth, differentiation and survival. However, mutant Ras-proteins that are always activated result in undifferentiated cell growth and are now implicated in greater than 30% of human cancers [111, 112]. The enzyme responsible for farnesylation (prenylation) of Ras, protein farnesyltransferase (PFTase), when inhibited resulted in reversion of the undifferentiated cell growth to a normal cell. Hence, synthesis of FPP analogs as anti-cancer therapeutics is an area of intense research [113]. Primary to this goal was the synthesis of suicide inhibitors against PFTase to prevent farnesylation through covalent attachment of the FPP analog to the enzyme by trapping a carbocation intermediate within the enzyme active site [114].

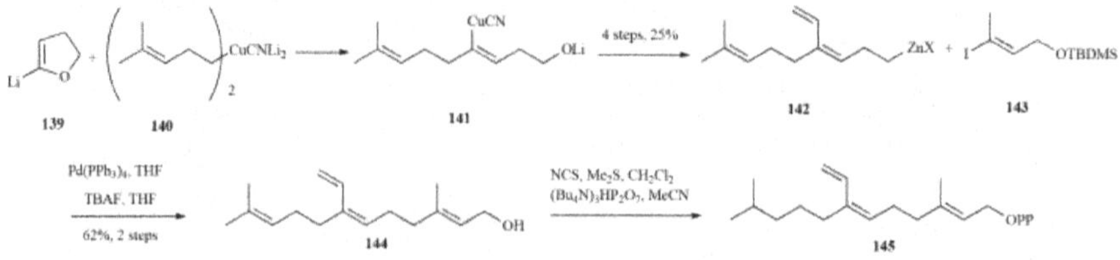

Scheme 29. Johnson's synthesis of a stereoisomeric mixture of 3-carene.

A primary concern in the synthesis of farnesyl diphosphate analogs is maintenance of the requisite all trans-isomer in the final product. Gibbs and co-workers designed multiple synthetic routes to both the 3- and 7-vinyl substituted farnesyl diphosphate (VFPP) analogs through the vinyl triflate intermediate 136 via a ketone ester 135 to provide the E-isomer 137 with high stereoselectivity (16:1) and in good yields [115]. Reduction, alcohol activation and diphosphorylation provided the all E-isomer of 3-VFPP 138 (Scheme 28). This synthetic approach was also implemented in the synthesis of the all E-7-VFPP analog; albeit, in much lower yields since two coupling steps using the vinyl triflate strategy were necessary, required 11 steps and less than 10% overall yield [116].

In a recent report, Gibbs et al. improved on their synthesis of 7-VFPP by coupling of 5-lithio-2, 3-dihydrofuran 139 with an isoprenylorganocuprate 140 to provide the corresponding alkenylcuprate intermediate 141. This compound proved to be a versatile intermediate since this affords the opportunity to introduce multiple functional groups into ultimately the 7-position of FPP. Following conversion of 141 into the corresponding organozinc intermediate 142 over four steps in 25% yield, the vinyl-substituted farnesyl skeleton was completed via coupling to the silyl-protected vinyl iodide 143 through palladium triphenylphosphine and removal of the silyl-protecting group to provide 7-vinylfarnesol 144. Diphosphorylation provided 7-VFPP 145 over six steps in 12% overall yield (Scheme 29) [117].

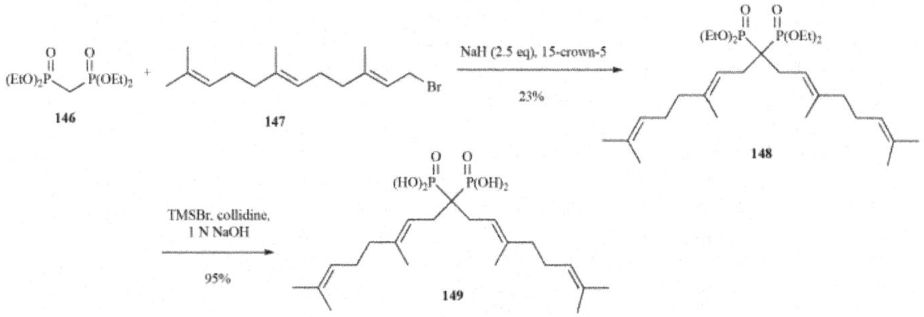

Scheme 30. Schull's synthesis of digeranyl-substituted bisphosphonate.

7.3.3 DITERPENES

Diterpenes, which comprise the C_{20} class of isoprenoids, are derived from condensation of IPP to FPP to form geranylgeranyl diphosphate (GGPP). Some representative examples include, dolichols, ubiquinones (plasto-quinones), phytyl derivatives and geranylgeranylated proteins [118]. The structural diversity is amplified through multivariant cyclizations via car-bocation intermediates and alkyl shifts to provide mono-, di-, and tricyclic compounds, which are further processed depending on organismal need. As a result of this diversity, it is not surprising that there are greater than 11,000 diterpenes that have been identified [119]. Therefore, in a similar vein as with the sesquiterpenes, this section will provide some representative examples for the chemical syntheses of GGPP analogs since there is ongoing interest for identifying novel compounds for medicinal purposes.

One class of inhibitors consists of the nitrogen-containing bisphospho-nates, which are used in treating bone resorption diseases such as osteo-porosis and metastatic bone disease [120]. These bisphosphonates work by depleting cells of GGPP, leading to inhibition of protein isoprenylation [121]. In order to determine the exact mechanism by which bisphospho-nates inhibit isoprenoid metabolism, a GGPP analog was synthesized using a synthetic scaffold that is amenable to varying isoprenoid chain lengths. A facile synthesis for one of the bisphosphonates containing the geranyl substituent is described below.

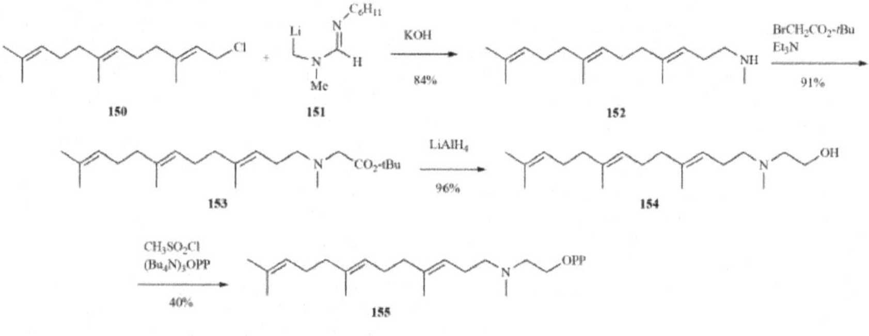

Scheme 31. Coates'synthesis of aza-GGPP.

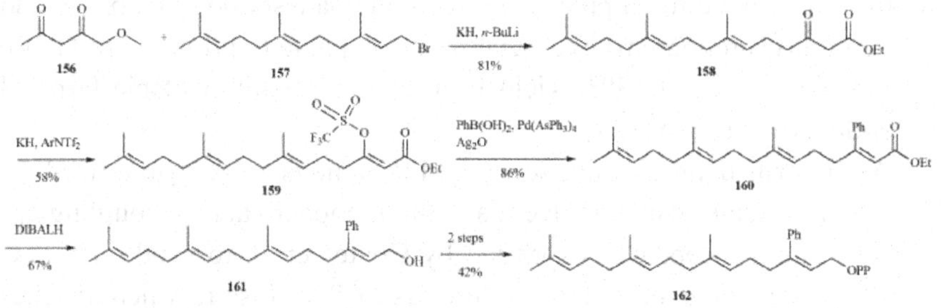

Scheme 32. Mu's synthesis of 3-PhGGPP.

Treatment of tetraethyl methylenebisphosphonate 146 with sodium hydride, 15-crown-5 ether, and farnesyl bromide 147 provided the di-alkylated bisphosphonate 148 product harboring the geranyl moieties. Although not high yielding, the mono- or dialkylation product could be regulated through careful titration with the base [122]. Deprotection of 148 with TMS-Br provided the GGPP analog 149 in 95% yield (Scheme 30). This strategy has proven successful for addition of alkylation reagents of varying chain length.

In addition to the bisphosphonates, another strategy using GGPP analogs was hypothesized for inhibition of GGPP synthase. One of the reasons for developing this alternative synthetic route was the observation that bisphosphonates tend to be poorly absorbed into the body through oral delivery. Creating compounds that have increased hydrophobicity was postulated to have better absorption properties [120]. Based upon biochemical evidence that a developing carbocation intermediate was formed within the enzyme active site of GGPP synthase the aza analog of GGPP was necessarily synthesized.

In the synthesis of an aza-GGPP 155, alkylation of farnesyl chloride 150 with α-lithio formamidine 151 generated an amidine, which was subsequently hydrolyzed to a farnesyl-N-methylamine 152. Alkylation of the amine with t-butylbromoacetate resulted in the β-aminoester 153 in excellent yield. Following reduction of 153 with lithium aluminum hydride, the resulting geranylgeraniol N-methylamine 154 was converted to the corresponding diphosphate with methanesulfonyl chloride and tris(tetrabutylammonium) hydrogen pyrophosphate to provide aza-GGPP 155 over four steps in 30% yield from commercially available farnesyl chloride (Scheme 31) [123].

As described previously, several FPP analogs were synthesized by Gibbs and Wiemer using a novel palladium, copper-catalyst coupling approach. This concept was expanded by Coates et al. to synthesize new GGPP analogs thought to act as inhibitors of PGGTase I, which attaches a geranylgeranyl moiety from GGPP to a cysteine in a CAAX-type box, where leucine is the carboxyl terminal residue [124]. Three new GGPP analogs were synthesized containing phenyl, tert-butyl and cyclopropyl substituents. The phenyl-substituted GGPP analog was prepared from an ethyl acetoacetate dianion 156 coupled with farnesyl bromide 157 to provide the β-ketoester 158. The potassium enolate was then transformed into the vinyl triflate 159. Coupling of the triflate with phenylboronic acid and silver oxide as the base yielded the desired phenyl ester 160. Reduction of the ester with DIBALH and subsequent phosphorylation provided 3-PhG-GPP 162 over six steps in 11% yield (Scheme 32) [125].

7.3.4 TRITERPENES

Triterpenes comprise the C_{30} class of isoprenoids and are biosynthesized through the condensation of two farnesyl diphosphate (FPP) units followed by reduction. The majority of triterpenes in biological systems result from 'head-to-head' condensations of FPP, which result in a linkage between

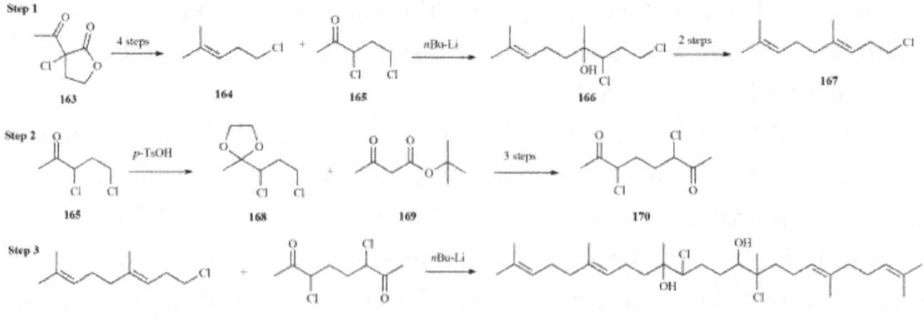

Scheme 33. Cornforth's synthesis of squalene.

the first carbons in each respective farnesyl group (a 1'-1 linkage). There are other examples of triterpenes with varied linkages in nature and one such triterpene, botryococcene (1'-3 linkage), will be discussed later in this section.

The linear triterpene, squalene, is a key precursor in the biosynthesis of cholesterol and other steroids. During the condensation of FPP monomers in the biosynthesis of squalene, a cyclopropyl diphosphate intermediate (presqualene diphosphate or PSPP) is formed and subsequently reduced by NADPH to form the final product [126]. Squalene itself then is transformed into the fused tetracyclic core characteristic of sterols and steroids. The synthesis of sterols, steroids, and other cyclic triterpenes represents a large body of information and is beyond the scope of this review. In this section, we will summarize approaches to construction of several representative linear triterpenes. Readers with an interest in cyclic triterpenes are directed to other comprehensive reviews.

An early synthesis of squalene was based on systematic olefin synthesis. In step 1, The five-carbon and six-carbon building blocks for the external segments of squalene were derived from α-acetyl-α-chlorobutyrolactone 163 to generate homoallylic chloride 164 and 3,5-dichloropentan-2-one 165 over four steps. Coupling of the C-5 and C-6 fragments with n-BuLi provided the homogeranyl backbone 166. Following epoxidation and

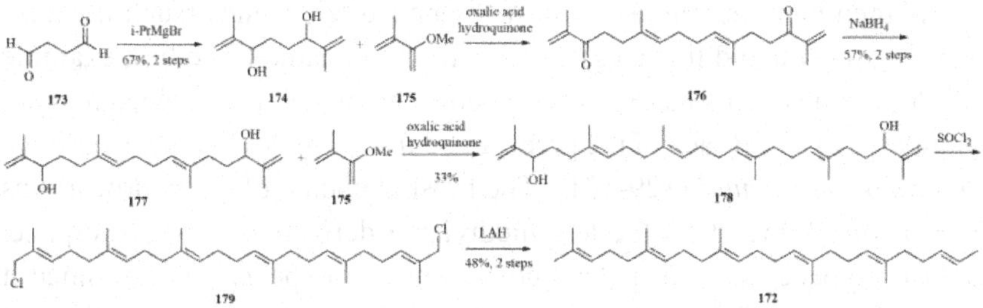

Scheme 34. Petersen's synthesis of squalene using a Claisen rearrangement.

elimination, the resulting homogeranyl chloride 167 was ready for squalene synthesis.

Step 2 entailed utilizing ketone 165 as the starting material for the synthesis of the internal segment of squalene. Briefly, following ketalization of 165 to generate 1, 3-dioxolane 168, coupling with t-butyl acetoacetate 169 and deprotection provided the central segment, 3,6-dichlorooctane-2,7-dione 170, for squalene assembly. Condensation of the products from steps 1 and 2 were accomplished by treatment of two equivalents of homogeranyl chloride 167 with two equivalents of n-BuLi and one equivalent of dione 170 to provide the dicholorosqualene intermediate 171. Subsequent epoxidation and oxygen elimination as employed in the construction of homogeranyl chloride provided squalene 172 in 15 steps (Scheme 33); however, yields for this synthesis were not estimated [127].

One challenge in the chemical synthesis of squalene is stereoselectivity, as all bonds are found in the trans-configuration. In a later synthesis, a stereoselective Claisen rearrangement was employed. Starting with succinaldehyde 173, treatment of the resulting alcohol with iPrMgBr yielded bis-allylic alcohol 174 in good yield. The alcohol was subject to a Claisen rearrangement with 3-methoxyisoprene 175 to the C-20 tetraenedione 176, which was subsequently reduced with sodium borohydride to provide the tetraenediol 177. Repetition of this sequence of reactions provided the C-30 squalene 178 backbone. Rearrangement of the terminal double bonds and elimination of the alcohol groups were accomplished through the use of thionyl chloride in a substitution reaction to provide the terminal halides 179, whereupon reduction with lithium aluminum hydride yielded squalene 172 over eight steps in 7% yield (Scheme 34) [128].

As mentioned previously, linear, branched triterpenes (such as those with 1'-3 isoprenoid linkages) are also found in nature. The best example of a triterpene of this nature is botryococcene, which is produced and accumulated by the B race of the green algae from which its name is derived, *Botryococcus braunii* [129-131]. The most abundant of the hydrocarbons in *B. braunii* Race B is the tetramethylated derivative of the triterpene, C-34 botryococcene [132]. This compound is of special interest since it was demonstrated the botryococcenes are compatible with existing oil

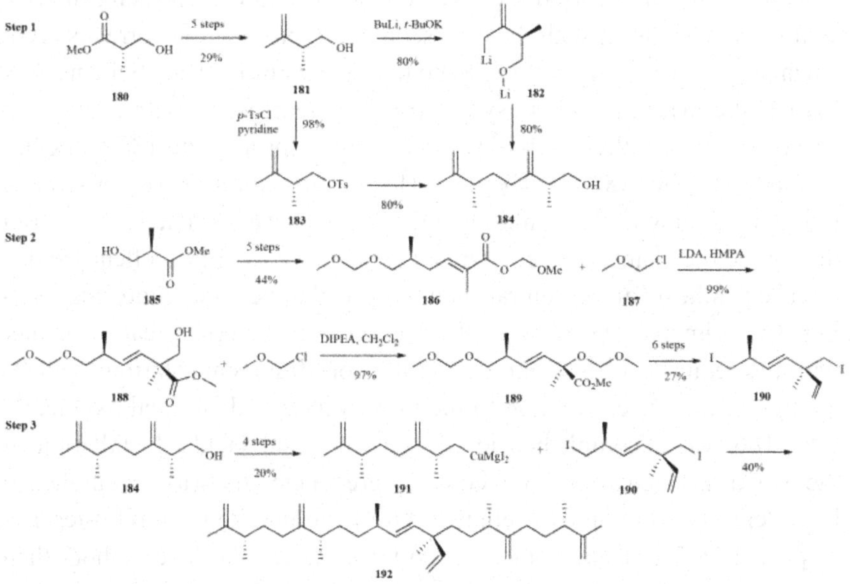

Scheme 35. White's synthesis of C34 botrycoccene.

infrastructures and, following hydrocracking, can be used directly as a 'drop-in' fuel [133].

The chemical synthesis of the tetramethylated botryococcene is challenging since there are six stereocenters on the carbon framework. To the best of our knowledge, there is only a single published report for the total synthesis of this isoprenoid with the correct stereochemistry in the final product, although there are several reports of the C-30 botryococcene and related analogs chemical synthesis [134-137]. Certainly, considering the interest in botryococcenes as a source of biofuel, novel synthetic approaches to the methylated versions may prove fruitful in order to study the enzyme(s) responsible for methylated botryococcene biosynthesis. Briefly, White et al. recognized the symmetry elements present in the structure and thus devised a convergent strategy for condensing the two symmetric fragments with the core segment to provide the correct stereochemistry in the final product. Specifically, as outlined in (Scheme 35), Step 1, the synthesis of the symmetric fragment began with a five step approach from methyl (2S)-3-hydroxy-2-methylpropionate 180 to the homoallylic alcohol 181 in 29% yield. Head-to-tail union of 181 proceeded through coupling of the dilithio species 182 with the tosylated alcohol 183 to provide the enantiomerically pure C_{12} segment 184 in excellent yield.

The synthesis of the central subunit proved to be much more challenging. The primary concern was introduction of a quaternary carbon center with the requisite R-configuration in the core fragment. Starting with R-methyl ester 185, conversion to the pivotal intermediate methoxymethyl ester 186 was accomplished in six steps in good yields. Installation of the quaternary center with the correct stereochemistry was accomplished by γ-deprotonation and 1,4-elimination of the methoxymethyl ester 187 to provide a 3:2 diastereomeric mixture of ester 188. The authors then used an enantioconvergent strategy to convert both diastereomeric intermediates to the central fragment through alternating protection and deprotection strategies to provide the enantiomerically pure diiodide central subunit 190 over thirteen steps in 7% yield. Activation of homoallylic alcohol 184 required four steps to the cuprate 191 intermediate and enabled

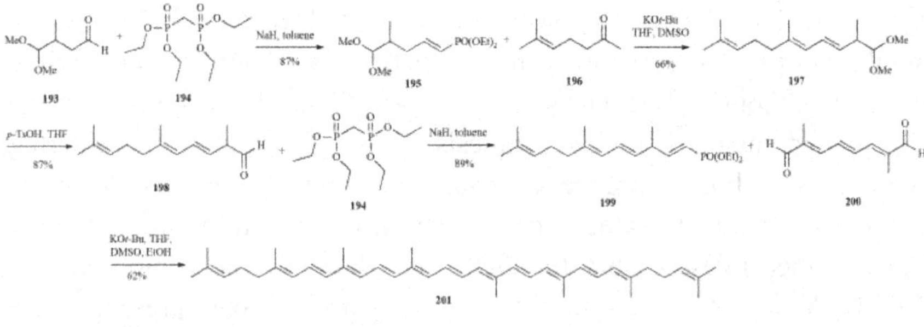

Scheme 36. Shen's synthesis of lycopene.

coupling of the core fragment 190 to provide $(-)$-C_{34}-botryococcene 192 over five steps in 8% yield (Scheme 35) [138].

7.3.5 TETRATERPENES

Tetraterpenes, which comprise the C_{40} class of isoprenoids, are derived from two C_{20} geranylgeranyl disphosphates in a head-to-head condensation reaction. Representative examples include the carotenoids and xanthophylls, which have strong anti-oxidant properties and are essential components in photosynthetic machinery [139]. Further, carotenoids are used extensively in the food and cosmetic industries as food colorants, protection against UV-irradiation, and as nutritional supplements [140, 141]. Although the majority of tetraterpenes are isolated from biological sources, the structural complexity of these isoprenoids proved to be a synthetic challenge largely due to the extensive double bond network and introduction of the correct stereochemistry at remote positions along the carbon framework. Two representative examples for achieving this end are presented.

Lycopene is an acyclic tetraterpene and the immediate precursor to both α- and β-carotene that consists of 13 double bonds all in the

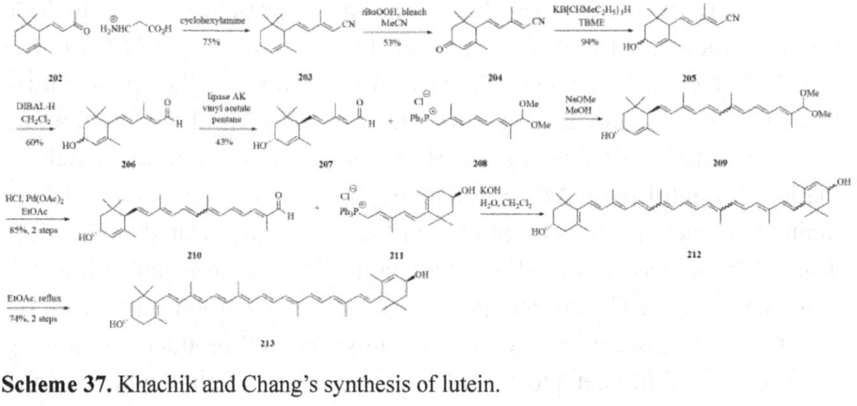

Scheme 37. Khachik and Chang's synthesis of lutein.

trans-configuration. This compound's high radical scavenging ability in addition to being synthetically challenging were major drivers toward lycopene synthesis [142]. A favorable synthetic route to the all trans-lycopene exploited two Wittig-Horner reactions to introduce double bonds under mild reaction conditions. The starting material, 4,4-dimethoxy-3-methylbutanal 193, was condensed with methylene bisphosphonic acid tetraethyl ester 194 in the presence of NaH to provide the C_6 propenyl phosphonate 195. Wittig-Horner condensation with the methyl heptenone 196 yielded the dimethoxytriene 197 in the correct double bond configuration. Following deprotection under acidic conditions, aldehyde 198 was condensed with the identical biphosphonic acid 194 to provide the all-E-tetraene 199. Another Wittig-Horner reaction with diketone 200 provided the all trans-lycopene 201 over five steps in 27% yield (Scheme 36) [143].

Lutein is a xanthophyll found in plants for non-photochemi-cal quenching of singlet excited state chlorophyll, which occurs at high light intensities, thus prevents photoinhibition. In addition, lutein is a major dietary carotenoid abundant in fruits and vegetables. The predominant stereoisomer is (3R,3'R,6'R)-β,ε-carotene-3,3'-diol [144]. The strategy used in one synthesis to make this isomer (and others) was to couple C-15 and C-10 building blocks, then elongate the resulting C-15 hydroxyaldehyde through a Wittig salt intermediate. Commercially available α-ionone 202 was the starting material for the synthesis of nitrile 203, which enabled facile chain elongation. Selective oxidation of 203 to the corresponding conjugated ketone 204 was accomplished in the presence of peroxide and bleach. Careful reduction of the ketone in the presence of borohydride provided racemic alcohol 205 in excellent yield. Reduction of the nitrile moiety to the aldehyde 206 using DIBAL-H resulted in the racemic mixture of hydroxyaldehydes. To isolate the desired (3R, 6R) configuration, kinetic resolution of 206 was employed using an enzyme-mediated acylation through Lipase AK. Fortunately, the desired stereoisomer 207 remained unreactive and was readily purified from the acylated enantiomer. The aldehyde was subjected to olefination using the protected Wittig salt 208 harboring the C_{10} internal portion of the lutein carbon framework resulting in a mixture of cis- and trans-configured 209 products. Following deprotection of the acetal to provide aldehyde 210, coupling with a second

Wittig salt reagent 211 appended to the remaining lutein framework provided a mixture of cis- and trans-lutein 212. The diastereomeric lutein was isomerized under refluxing conditions and isolated via column chromatography and crystallization resulting in stereomerically pure (3R,3'R,6'R) lutein 213 in 6% yield over nine steps (Scheme 37) [145].

7.4 SUMMARY

Isoprenoids are a tremendous illustration of the cornucopia of chemical diversity and unique biochemical roles that are possible within members of a single molecular family. A detailed understanding of these structures, and of their roles, is empowered by the development of synthetic methodologies to produce the corresponding target molecules, and as substrates or products for detailed enzymological studies. In nature, the biosynthesis of isoprenoids affects the conversion of small molecule acids (such as those used in the mevalonate pathway) or carbohydrates (such as those used in the methylerythritol phosphate pathway) into molecules that are largely aliphatic hydrocarbons. Similarly, as is outlined in our review, the synthetic methodologies to produce a given isoprenoid may draw from and utilize chemical reactions associated with the construction of hydrocarbons, carbohydrates, or diphosphate based bioconjugates.

Whereas the biological production of isoprenoids will likely be the major route to these compounds, chemical synthesis of isoprenoids and their corresponding precursors will always be necessary. This is especially relevant when probing enzyme activity thus requiring the synthesis of substrate analogs. Emphasis on the chemical syntheses of MEP pathway intermediates to isoprenoids was a result of the exhaustive efforts required to probe the enzyme function responsible for tens of thousands of natural products that are found in all corners of life. It is also important to highlight the potential of longer chain isoprenoids for use in industrial applications, including nutracueticals, pharmaceuticals, cosmetics and as a source for biofuels due to the energy rich composition of these important compounds.

REFERENCES

1. Christianson DW. Unearthing the roots of the terpenome. Curr. Opin. Chem. Biol. 2008;12:141–150.
2. Bach TJ, Rohmer M. Isoprenoid Synthesis in Plants and Microorganisms. Springer: New York. 2012:505.
3. Chappell J. Biochemistry and molecular biology of the isoprenoid biosynthetic pathway in plants. Annu. Rev. Plant. Phys. 1995;46:521–547.
4. Kirby J, Keasling JD. Biosynthesis of plant isoprenoids: Perspectives for microbial engineering. Annu. Rev. Plant. Biol. 2009;60:335–355.
5. Eubanks LM, Poulter CD. Rhodobacter capsulatus DXP synthase: Steady-state kinetics and substrate binding. Biochemistry. 2001;40:8613–8614.
6. Koppisch AT, Fox DT, Blagg BS, Poulter CD. E.coli MEP synthase: steady-state kinetic analysis and substrate binding. Biochemistry. 2002;41:236–243.
7. Rohdich F, Schuhr CA, Hecht S, Herz S, Wungsintaweekul J, Eisenreich W, Zenk MH, Bacher A. Biosynthesis of isoprenoids.A rapid method for the preparation of isotope-labeled 4-diphosphocytidyl-2C-methyl-D-erythritol. J. Am. Chem. Soc. 2000;122:9571–9574.
8. Miallau L, Alphey MS, Kemp LE, Leonard GA, McSweeney SM, Hecht S, Bacher A, Eisenreich W, Rohdich F, Hunter WN. Biosynthesis of isoprenoids: crystal structure of 4-diphosphocytidyl-2C-methyl-D-erythritol kinase. Proc. Natl. Acad. Sci. USA. 2003;100:9173–9178.
9. Herz S, Wungsintaweekul J, Schuhr CA, Hecht S, Luttgen H, Sagner S, Fellermeier M, Eisenreich W, Zenk MH, Bacher A, Rohdich F. Biosynthesis of terpenoids: YgbB protein converts 4-diphosphocytidyl-2C-methyl-D-erythritol 2-phosphate to 2C-methyl-D-erythritol 2, 4-cyclo-diphosphate. Proc. Natl. Acad. Sci. USA. 2000;97:2486–2490.
10. Seemann M, Rohmer M. Isoprenoid biosynthesis via the methylerythritol phosphate pathway: GcpE and LytB, two novel iron-sulphur proteins. Cr. Chim. 2007;10:748–755.
11. Zhou CF, Li ZR, Wiberley-Bradford AE, Weise SE, Sharkey TD. Isopentenyl diphosphate and dimethylallyl diphosphate/isopentenyl diphosphate ratio measured with recombinant isopentenyl diphosphate isomerase and isoprene synthase. Anal. Biochem. 2013;440:130–136.
12. Eisenreich W, Schwarz M, Cartayrade A, Arigoni D, Zenk MH, Bacher A. The deoxyxylulose phosphate pathway of terpenoid biosynthesis in plants and microorganisms. Chem. Biol. 1998;5:R221–233.
13. Rohmer M. The discovery of a mevalonate-independent pathway for isoprenoid biosynthesis in bacteria, algae and higher plants. Nat. Prod. Rep. 1999;16:565–74.
14. Jomaa H, Wiesner J, Sanderbrand S, Altincicek B, Weidemeyer C, Hintz M, Turbachova I, Eberl M, Zeidler J, Lichtenthaler HK, Soldati D, Beck E. Inhibitors of the nonmevalonate pathway of isoprenoid biosynthesis as antimalarial drugs. Science. 1999;285:1573–1576.
15. Testa CA, Brown MJ. The methylerythritol phosphate pathway and its significance as a novel drug target. Curr. Pharm. Biotech. 2003;4:248–259.

16. Putra SR, Lois LM, Campos N, Boronat A, Rohmer M. Incorporation of [2,3-C-13(2)]- and [2,4-C-13(2)]-D-1-deoxyxylulose into ubiquinone of Escherichia coli via the mevalonate-independent pathway for isoprenoid biosynthesis. Tetrahedron Lett. 1998;39:23–26.

17. Shabat D, List B, Lerner RA, Barbas CF. A short enantioselective synthesis of 1-deoxy-L-xylulose by antibody catalysis. Tetrahedron Lett. 1999;40:1437–1440.

18. Hecht S, Kis F, Eisenreich W, Amslinger S, Wungsintaweekul J, Herz S, Rohdich F, Bacher A. Enzyme-assisted preparation of isotope-labeled 1-deoxy-D-xylulose 5-phosphate. J. Org. Chem. 2001;66:3948–3952.

19. Zhou YF, Cui Z, Li H, Tian J, Gao WY. Optimized enzymatic preparation of 1-deoxy-D-xylulose 5-phosphate. Bioorg. Chem. 2010;38:120–123.

20. Therisod M, Fischer JC, Estramareix B. The origin of the carbon chain in the thiazole moiety of thiamine in Escherichia coli - Incorporation of deuterated 1-deoxy-D-threo-2-pentulose. Biochem. Biophys. Res. Commun. 1981;98:374–379.

21. David S, Estramareix B, Fischer JC, Therisod M. 1-Deoxy-D-threo-2-pentulose - the precursor of the 5-carbon chain of the thiazole of thiamine. J. Am. Chem. Soc. 1981;103:7341–7342.

22. David S, Estramareix B, Fischer JC, Therisod M. The biosynthesis of thiamine - syntheses of [1,1,1,5-(H4)-H-2]-1-deoxy-D-threo-2-pentulose and incorporation of this sugar in biosynthesis of thiazole by Escherichia coli cells. J. Chem. Soc. Perk. T 1. 1982:2131–2137.

23. Kennedy IA, Hill RE, Pauloski RM, Sayer BG, Spenser ID. Biosynthesis of vitamin B-6 - Origin of pyridoxine by the union of 2 acyclic precursors, 1-deoxy-D-xylulose and 4-hydroxy-L-threonine. J. Am. Chem. Soc. 1995;117:1661–1662.

24. Giner JL, Jaun B. Biosynthesis of isoprenoids in Escherichia coli: Retention of the methyl H-atoms of 1-deoxy-D-xylulose. Tetrahedron Lett. 1998;39:8021–8022.

25. Meyer O, Hoeffler JF, Grosdemange-Billiard C, Rohmer M. Practical synthesis of 1-deoxy-D-xylulose and 1-deoxy-D-xylulose 5-phosphate allowing deuterium labelling. Tetrahedron. 2004;60:12153–12162.

26. Blagg BSJ, Poulter CD. Synthesis of 1-deoxy-D-xylulose and 1-deoxy-D-xylulose-5-phosphate. J. Org. Chem. 1999;64:1508–1511.

27. Crispino GA, Ho PT, Sharpless KB. Selective perhydroxylation of squalene - taming the arithmetic demon. Science. 1993;259:64–66.

28. Wang ZM, Sharpless KB. Asymmetric dihydroxylation of alpha-substituted styrene derivatives. Synlett (11): 1993:603–604.

29. Giner JL, Jaun B, Arigoni D. Biosynthesis of isoprenoids in Escherichia coli: The fate of the 3-H and 4-H atoms of 1-deoxy-d-xylulose. Chem. Commun. (Camb) 1998:1857–1858.

30. Piel J, Boland W. Highly efficient and versatile synthesis of isotopically labelled 1-deoxy-D-xylulose. Tetrahedron Lett. 1997;38:6387–6390.

31. Taylor SV, Vu LD, Begley TP, Schorken U, Grolle S, Sprenger GA, Bringer-Meyer S, Sahm H. Chemical and enzymatic synthesis of 1-Deoxy-D-xylulose-5-phosphate. J. Org. Chem. 1998;63:2375–2377.

32. Giner JL. New and efficient synthetic routes to 1-deoxy-D-xylulose. Tetrahedron Lett. 1998;39:2479–2482.

33. Cox RJ, de Andres-Gomez A, Godfrey CRA. Rapid and flexible synthesis of 1-deoxy-D-xylulose-5-phosphate, the substrate for 1-deoxy-D-xylulose-5-phosphate reductoisomerase. Org. Biomol. Chem. 2003;1:3173–3177.

34. Wong U, Cox RJ. The chemical mechanism of D-1-deoxyxylulose-5-phosphate reductoisomerase from Escherichia coli. Angew. Chem. Int. Ed. Engl. 2007;46:4926–4929.

35. Fox DT, Poulter CD. Mechanistic studies with 2-C-methyl-D-erythritol 4-phosphate synthase from Escherichia coli. Biochemistry. 2005;44:8360–8368.

36. Wong A, Munos JW, Devasthali V, Johnson KA, Liu HW. Study of 1-deoxy-D-xylulose-5-phosphate reductoisomerase: Synthesis and evaluation of fluorinated substrate analogues. Org. Lett. 2004;6:3625–3628.

37. Munos JW, Pu XT, Mansoorabadi SO, Kim HJ, Liu HW. A secondary kinetic isotope effect study of the 1-deoxy-D-xylulose-5-phosphate reductoisomerase-catalyzed reaction: Evidence for a retroaldol-aldol rearrangement. J. Am. Chem. Soc. 2009;131:2048–2049.

38. Bouvet D, O'Hagan D. The synthesis of 1-fluoro- and 1,1-difluoro- analogues of 1-deoxy-D-xylulose. Tetrahedron. 1999;55:10481–10486.

39. Fox DT, Poulter CD. Synthesis and evaluation of 1-deoxy-D-xylulose 5-phosphoric acid analogues as alternate substrates for methylerythritol phosphate synthase. J. Org. Chem. 2005;70:1978–1985.

40. Muehlbacher M, Poulter CD. Isopentenyl-diphosphate isomerase: inactivation of the enzyme with active-site-directed irreversible inhibitors and transition-state analogues. Biochemistry. 1988;27:7315–7328.

41. Prakash GKS, Krishnamurti R, Olah GA. Synthetic methods and reactions.141. Fluoride-induced trifluoromethylation of carbonyl-compounds with trifluoromethyltrimethylsilane (TMS-CF3) - a trifluoro-methide equivalent. J. Am. Chem. Soc. 1989;111:393–395.

42. Testa CA, Cornish RA, Poulter CD. The sorbitol phosphotransferase system is responsible for transport of 2-C-methyl-D-erythritol into Salmonella enterica serovar typhimurium. J. Bacteriol. 2004;186:473–480.

43. Duvold T, Cali P, Bravo JM, Rohmer M. Incorporation of 2-C-methyl-D-erythritol, a putative isoprenoid precursor in the mevalonate-independent pathway, into ubiquinone and menaquinone of Escherichia coli. Tetrahedron Lett. 1997;38:6181–6184.

44. Charon L, Hoeffler JF, Pale-Grosdemange C, Lois LM, Campos N, Boronat A, Rohmer M. Deuterium-labelled isotopomers of 2-C-methyl-D-erythritol as tools for the elucidation of the 2-C-methyl-D-erythritol 4-phosphate pathway for isoprenoid biosynthesis. Biochem. J. 2000;346 (Pt 3):737–42.

45. Charon L, Hoeffler JF, Pale-Grosdemange C, Rohmer M. Synthesis of [3,5,5,5-H-2(4)]-2-C-methyl-D-erythritol, a substrate designed for the elucidation of the mevalonate independent route for isoprenoid biosynthesis. Tetrahedron Lett. 1999;40:8369–8373.

46. Bitok JK, Meyers CF. 2C-Methyl-D-erythritol 4-phosphate enhances and sustains cyclodiphosphate synthase IspF activity. ACS Chem. Biol. 2012;7:1702–1710.

47. Hemmerlin A, Hoeffler JF, Meyer O, Tritsch D, Kagan IA, Grosdemange-Billiard C, Rohmer M, Bach TJ. Cross-talk between the cytosolic mevalonate and the plastidial methylerythritol phosphate pathways in Tobacco Bright Yellow-2 cells. J. Biol. Chem. 2003;278:26666–26676.
48. Kuzuyama T, Takahashi S, Watanabe H, Seto H. Direct formation of 2-C-methyl-D-erythritol 4-phosphate from 1-deoxy-D-xylulose 5-phosphate by 1-deoxy-D-xylulose 5-phosphate reductoisomerase, a new enzyme in the non-mevalonate pathway to isopentenyl diphosphate. Tetrahedron Lett. 1998;39:4509–4512.
49. Sagner S, Eisenreich W, Fellermeier M, Latzel C, Bacher A, Zenk MH. Biosynthesis of 2-C-methyl-D-erythritol in plants by rearrangement of the terpenoid precursor, 1-deoxy-D-xylulose 5-phosphate. Tetrahedron Lett. 1998;39:2091–2094.
50. Ding X, Zheng M, Yu LP, Zhang XL, Weber RJ, Yan B, Russell AG, Edgerton ES, Wang XM. Spatial and seasonal trends in biogenic secondary organic aerosol tracers and water-soluble organic carbon in the southeastern United States. Environ. Sci. Technol. 2008;42:5171–5176.
51. Li QA, Wang W, Zhang HW, Wang YJ, Wang B, Li L, Li HJ, Wang BJ, Zhan J, Wu M, Bi XH. Development of a compound-specific carbon isotope analysis method for 2-methyltetrols, biomarkers for secondary organic aerosols from atmospheric isoprene. Anal. Chem. 2010;82:6764–6769.
52. Xia X, Hopke PK. Seasonal variation of 2-methyltetrols in ambient air samples. Environ. Sci. Technol. 2006;40:6934–6937.
53. Witczak ZJ, Whistler RL. Synthesis of 3-C-(hydroxymethyl)erythritol and 3-C-methylerythritol. Carbohydr. Res. 1984;133: 235–245.
54. Duvold T, Bravo JM, PaleGrosdemange C, Rohmer M. Biosynthesis of 2-C-methyl-D-erythritol, a putative C-5 intermediate in the mevalonate independent pathway for isoprenoid biosynthesis. Tetrahedron Lett. 1997;38:4769–4772.
55. Kolb HC, Vannieuwenhze MS, Sharpless KB. Catalytic asymmetric dihydroxylation. Chem. Rev. 1994;94:2483–2547.
56. Fontana A, Messina R, Spinella A, Cimino G. Simple and versatile synthesis of branched polyols: (+)-2-C-methylerythritol and (+)-2-C-methylthreitol. Tetrahedron Lett. 2000;41:7559–7562.
57. Wang H, Zhao XM, Li YH, Lu L. Asymmetric synthesis of four isomers of 2-C-trifluoromethylerythritol. J. Org. Chem. 2006;71:3278–3281.
58. Ghosh SK, Butler MS, Lear MJ. Synthesis of 2-C-methylerythritols and 2-C-methylthreitols via enantiodivergent sharp less dihydroxylation of trisubstituted olefins. Tetrahedron Lett. 2012;53:2706–2708.
59. Koppisch AT, Blagg BSJ, Poulter CD. Synthesis of 2-C-methyl-D-erythritol 4-phosphate: The first pathway-specific intermediate in the methylerythritol phosphate route to isoprenoids. Org. Lett. 2000;2:215–217.
60. Koppisch AT, Poulter CD. Synthesis of 4-diphosphocytidyl-2-C-methyl-D-erythritol and 2-C-methyl-D-erythritol-4-phosphate. J. Org. Chem. 2002;67:5416–5418.
61. Robinson TV, Pedersen DS, Taylor DK, Tiekink ERT. Dihydroxylation of 4-substituted 1,2-dioxines: A concise route to branched erythro sugars. J. Org. Chem. 2009;74:5093–5096.

62. Kis K, Wungsintaweekul J, Eisenreich W, Zenk MH, Bacher A. An efficient preparation of 2-C-methyl-D-erythritol 4-phosphoric acid and its derivatives. J. Org. Chem. 2000;65:587–592.

63. Hoeffler JF, Grosdemange-Billiard C, Rohmer M. Synthesis of tritium labelled 2-C-methyl-D-erythritol, a useful substrate for the elucidation of the methylerythritol phosphate pathway for isoprenoid biosynthesis. Tetrahedron Lett. 2000;41:4885–4889.

64. Hoeffler JF, Pale-Grosdemange C, Rohmer M. Chemical synthesis of enantiopure 2-C-methyl-D-erythritol 4-phosphate, the key intermediate in the mevalonate-independent pathway for isoprenoid biosynthesis. Tetrahedron. 2000;56:1485–1489.

65. Lagisetti C, Urbansky M, Coates RM. The dioxanone approach to (2S 3R)-2-C-methylerythritol 4-phosphate and 2-4-cyclodiphosphate and various MEP analogues. J. Org Chem. 2007;72:9886–9895.

66. Urbansky M, Davis CE, Surjan JD, Coates RM. Synthesis of enantiopure 2-C-methyl-D-erythritol 4-phosphate and 2-4-cyclodiphosphate from D-arabitol. Org. Lett. 2004;6:135–138.

67. Odejinmi SI, Rascon RG, Chen W, Lai K. Formal synthesis of 4-diphosphocytidyl-2-C-methyl D-erythritol from D-(+)-arabitol. Tetrahedron. 2012;68:8937–8941.

68. Sharma A, Das P, Chattopadhyay S. Concise asymmetric syntheses of the (+)-2-C-methyltetritol isomers. Tetrahedron Asymmetry. 2008;19:02167–2170.

69. Koppisch AT, Blagg BS, Poulter CD. Synthesis of 2-C-methyl-D-erythritol 4-phosphate the first pathway-specific intermediate in the methylerythritol phosphate route to isoprenoids. Org Lett. 2000;2:215–217.

70. Bennani YL, Sharpless KB. Asymmetric dihydroxylation (Ad) of NN-DIALKYL and N-methoxy-N-methyl alpha beta-unsaturated and beta gamma-unsaturated amides. Tetrahedron Lett. 1993;34:2079–2082.

71. Fontana A. Concise synthesis of (+)-2-C-methyl-D-erythritol-4-phosphate. J. Org. Chem. 2001;66:2506–2508.

72. Kis K, Wungsintaweekul J, Eisenreich W, Zenk MH, Bacher A. An efficient preparation of 2-C-methyl-D-erythritol 4-phosphoric acid and its derivatives. J. Org. Chem. 2000;65:587–592.

73. Hirsch G, Grosdemange-Billiard C, Tritsch D, Rohmer M. (3R 4S)-3 4 5-trihydroxy-4-methylpentylphosphonic acid, an isosteric phosphonate analogue of 2-C-methyl-D-erythritol 4-phosphate, a key intermediate in the new pathway for isoprenoid biosynthesis. Tetrahedron Lett. 2004;45:519–521.

74. Koumbis AE, Kotoulas SS, Gallos JK. A convenient synthesis of 2-C-methyl-D-erythritol 4-phosphate and isotopomers of its precursor. Tetrahedron. 2007;63:2235–2243.

75. David S, Hanessian S. Regioselective manipulation of hydroxyl groups via organotin derivatives. Tetrahedron. 1985;41:0643–663.

76. Narayanasamy P, Eoh H, Brennan PJ, Crick DC. Synthesis of 4-diphosphocytidyl-2-C-methyl-D-erythritol 2-phosphate and kinetic studies of Mycobacterium tuberculosis IspF. Chem. Biol. 2010;17:117–122.

77. Narayanasamy P, Eoh H, Crick DC. Chemoenzymatic synthesis of 4-diphosphocytidyl-2-C-methyl-D-erythritol: a substrate for IspE. Tetrahedron Lett. 2008;49:4461–4463.

78. Marlow AL, Kiessling LL. Improved chemical synthesis of UDP-galactofuranose. Org. Lett. 2001;3:2517–2519.

79. Giner JL, Ferris WV. Synthesis of 2-C-methyl-D-erythritol 2,4-cyclopyrophosphate. Org. Lett. 2002;4:1225–1226.

80. Fox DT, Poulter CD. Synthesis of (E)-4-hydroxydimethylallyl diphosphate.An intermediate in the methyl erythritol phosphate branch of the isoprenoid pathway. J. Org. Chem. 2002;67:5009–5010.

81. Amslinger S, Kis K, Hecht S, Adam P, Rohdich F, Arigoni D, Bacher A, Eisenreich W. Biosynthesis of terpenes.preparation of (E)-1-Hydroxy-2-methyl-but-2-enyl 4-diphoshate an intermediate of the deoxyxylulose phosphate pathway. J. Org. Chem. 2002; 67:4590–4594.

82. Ward JL, Beale MH. Synthesis of (2E)-4-hydroxy-3-methylbut-2-enyl diphosphate, a key intermediate in the biosynthesis of isoprenoids. J. Chem. Soc. Perk. T. 1. 2002:710–712.

83. Giner JL. A convenient synthesis of (E)-4-hydroxy-3-methyl-2-butenyl pyro-phosphate and its [4-C-13]-labeled form. Tetrahedron Lett. 2002;43:5457–5459.

84. Gao WY, Loeser R, Raschke M, Dessoy MA, Fulhorst M, Alpermann H, Wessjohann LA, Zenk MH. (E)-4-hydroxy-3-methylbut-2-enyl diphosphate: An intermediate in the formation of terpenoids in plant chromoplasts. Angew. Chem. Int. Ed. Engl. 2002;41:2604–2607.

85. Hecht S, Amslinger S, Jauch J, Kis K, Trentinaglia V, Adam P, Eisenreich W, Bacher A, Rohdich F. Studies on the non-mevalonate isoprenoid biosynthetic pathway.Simple methods for preparation of isotope-labeled (E)-1-hydroxy-2-methylbut-2-enyl 4-diphosphate. Tetrahedron Lett. 2002;43:8929–8933.

86. Wolff M, Seemann M, Grosdemange-Billiard C, Tritsch D, Campos N, Rodriguez-Concepcion M, Boronat A, Rohmer M. Isoprenoid biosynthesis via the methylerythritol phosphate pathway.(E)-4-hydroxy-3-methylbut-2-enyl diphosphate: chemical synthesis and formation from methylerythritol cyclodiphosphate by a cell-free system from Escherichia coli. Tetrahedron Lett. 2002;43:2555–2559.

87. Fox DT, Poulter CD. Synthesis of (E)-4-hydroxydimethylallyl diphosphate.An intermediate in the methyl erythritol phosphate branch of the isoprenoid pathway. J. Org. Chem. 2002;67:5009–5010.

88. Rohdich F, Hecht S, Gartner K, Adam P, Krieger C, Amslinger S, Arigoni D, Bacher A, Eisenreich W. Studies on the nonmevalonate terpene biosynthetic pathway: Metabolic role of IspH (LytB) protein. Proc. Natl. Acad. Sci. U S A. 2002;99:1158–1163.

89. Cramer F, Bohm W. Synthese von geranyl-pyrophospat und farnesyl-pyrophosphat. Angew. Chem. Int. Edit. 1959;71:775–775.

90. Davisson VJ, Woodside AB, Neal TR, Stremler KE, Muehlbacher M, Poulter CD. Phosphorylation of isoprenoid alcohols. J. Org. Chem. 1986;51:4768–4779.

91. Davisson VJ, Woodside AB, Poulter CD. Synthesis of allylic and homoallylic isoprenoid pyrophosphates. Method. Enzymol. 1985;110:130–144. [PubMed]

92. Walsh CT. Revealing coupling patterns in isoprenoid alkylation biocatalysis. ACS Chem. Biol. 2007;2:296–298.

93. Thulasiram HV, Erickson HK, Poulter CD. Chimeras of two isoprenoid synthases catalyze all four coupling reactions in isoprenoid biosynthesis. Science. 2007;316:73–76.

94. Thulasiram HV, Erickson HK, Poulter CD. A common mechanism for branching, cyclopropanation, and cyclobutanation reactions in the isoprenoid biosynthetic pathway. J. Am. Chem. Soc. 2008;130:1966–1971.

95. Mcgarvey DJ, Croteau R. Terpenoid metabolism. Plant Cell. 1995;7:1015–1026.

96. Lange BM, Ahkami A. Metabolic engineering of plant monoterpenes, sesquiterpenes and diterpenes-current status and future opportunities. Plant Biotechnol. J. 2013;11:169–196.

97. Thomas AF, Bessiere Y, ApSimon J, editors. 7 pp. John Wiley & Sons Inc Hoboken: In: The Total Synthesis of Natural Products ; 1988. The synthesis of monoterpenes 1980-1986. pp. 274–454.

98. Borschberg HR. New strategies for the synthesis of monoterpene indole alkaloids. Curr. Org. Chem. 2005;9:1465–1491.

99. Takacs JM, Boito SC, Myoung YC. Recent applications of catalytic metal-mediated carbocyclizations in asymmetric synthesis. Curr. Org. Chem. 1998;2:233–254.

100. Frauenfelder C, Schmid GA, Vogelsang T, Borschberg HJ. Flexible synthetic approaches to monoterpene indole alkaloids. Chimia. 2001;55:828–830.

101. Bettadaiah BK, editor. University of Msore Mysore: 2003. Studies on new synthetic strategies for O- and S-derivatives of monoterpenes.

102. Thomas AF, Bessiere Y. Limonene. Nat. Prod. Rep. 1989;6:291–309.

103. Money T. Camphor - A chiral starting material in natural product synthesis. Nat. Prod. Rep. 1985;2:253–289.

104. de Carvalho C, da Fonseca MMR. Biotransformation of terpenes. Biotechnol. Adv. 2006;24:134–142.

105. Thomas MT, Fallis AG. Total synthesis of (+/-) alpha-pinene and (+/-) beta-pinene - general route to bicyclo[3..1]heptanes. . Tetrahedron Lett. 1973:4687–4690.

106. Kulkarni YS, Snider BB. Intramolecular [2 + 2] cyclo-additions of ketenes2. Synthesis of chrysanth.none beta-pinene, beta-cis-bergamotene, and beta-trans-Bergamotene. J. Org. Chem . 1985;50:2809–2810.

107. Cocker W. Review of some investigations of chemistry of carene. J. Soc. Cosmet. Chem. 1971;22:249–284.

108. Kadow JF, Johnson CR. Synthesis of fused cyclopropanes from gamma-stannyl alcohols. Tetrahedron Lett. 1984;25:5255–5258.

109. Macaev FZ, Malkov AV. Use of monoterpenes, 3-carene and 2-carene, as synthons in the stereoselective synthesis of 2,2-dimethyl-1,3-disubstituted cyclopropanes. Tetrahedron. 2006;62:9–29.

110. Merfort I. Perspectives on sesquiterpene lactones in inflammation and cancer. Curr. Drug Targets. 2011;12:1560–1573.

111. Bos JL. Ras oncogenes in human cancer - a review. Cancer Res. 1989;49:4682–4689.

112. Harris CM, Poulter CD. Recent studies of the mechanism of protein prenylation. Nat. Prod. Rep. 2000;17:137–144.

113. Ohkanda J, Knowles DB, Blaskovich MA, Sebti SM, Hamilton AD. Inhibitors of protein farnesyltransferase as novel anticancer agents. Curr. Top. Med. Chem. 2002;2:303–323.

114. Mu YQ, Gibbs RA, Eubanks LM, Poulter CD. Cuprate-mediated synthesis and biological evaluation of cyclopropyl- and tert-butylfarnesyl diphosphate analogs. J. Org. Chem. 1996;61:8010–8015.

115. Gibbs RA, Krishnan U, Dolence JM, Poulter CD. A stereoselective palladium copper-catalyzed route to isoprenoids - synthesis and biological evaluation of 13-methylidenefarnesyl diphosphate. J. Org. Chem. 1995;60:7821–7829.

116. Rawat DS, Gibbs RA. Synthesis of 7-substituted farnesyl diphosphate analogues. Org. Lett. 2002;4:3027–3030.

117. Placzek AT, Gibbs RA. New synthetic methodology for the construction of 7-substituted farnesyl diphosphate analogs. Org. Lett. 2011;13:3576–3579.

118. Wiemer AJ, Wiemer DF, Hohl RJ. Geranylgeranyl diphosphate synthase: An emerging therapeutic target. Clin. Pharmacol. Ther. 2011;90:804–812.

119. Boca Raton Florida: 2003. Dictionary of Natural Products Chapman & Hall through CRC Press Version 11.

120. Szabo CM, Matsumura Y, Fukura S, Martin MB, Sanders JM, Sengupta S, Cieslak JA, Loftus TC, Lea CR, Lee HJ, Koohang A, Coates RM, Sagami H, Oldfield E. Inhibition of geranylgeranyl diphosphate synthase by bisphosphonates and diphosphates: A potential route to new bone antiresorption and antiparasitic agents. J. Med. Chem. 2002;45:2185–2196.

121. Wiemer AJ, Yu JS, Lamb KM, Hohl RJ, Wiemer DF. Mono- and dialkyl isoprenoid bisphosphonates as geranylgeranyl diphosphate synthase inhibitors. Bioorgan. Med. Chem. 2008;16:390–399.

122. Shull LW, Wiemer AJ, Hohl RJ, Wiemer DF. Synthesis and biological activity of isoprenoid bisphosphonates. Bioorg. Med. Chem. 2006;14:4130–4136.

123. Steiger A, Pyun HJ, Coates RM. Synthesis and characterization of aza analog inhibitors of squalene and geranylgeranyl diphosphate synthases. J. Org. Chem. 1992;57:3444–3449.

124. Casey PJ, Seabra MC. Protein prenyltransferases. J. Biol. Chem. 1996;271:5289–5292.

125. Mu YQ, Eubanks LM, Poulter CD, Gibbs RA. Coupling of isoprenoid triflates with organoboron nucleophiles: Synthesis and biological evaluation of geranylgeranyl diphosphate analogues. Bioorg. Med. Chem. 2002;10:1207–1219.

126. Dewar MJS, Ruiz JM. Mechanism of the biosynthesis of squalene from farnesyl pyrophosphate. Tetrahedron. 1987;43:2661–2674.

127. Cornforth JW, Cornforth RH, Mathew KK. A stereoselective synthesis of squalene. J. Chem. Soc. 1959:2539–2547.

128. Faulkner DJ, Petersen MR. Application of claisen rearrangement to synthesis of trans trisubstituted olefinic bonds - synthesis of squalene and insect juvenile-hormone. J. Am. Chem. Soc. 1973;95:553–563.

129. Weiss TL, Roth R, Goodson C, Vitha S, Black I, Azadi P, Rusch J, Holzenburg A, Devarenne TP, Goodenough U. Colony organization in the green alga Botryococcus braunii (Race B) is specified by a complex extracellular matrix. Eukaryot. Cell. 2012;11:1424–1440.

130. Eroglu E, Okada S, Melis A. Hydrocarbon productivities in different Botryococcus strains: comparative methods in product quantification. J. Appl. Phycol. 2011;23:763–775.

131. Lovejoy KS, Davis LE, McClellan LM, Lillo AM, Welsh JD, Schmidt EN, Sanders CK, Lou AJ, Fox DT, Koppisch AT, Del Sesto RE. Evaluation of ionic liquids on phototrophic microbes and their use in biofuel extraction and isolation. J. Appl. Phycol. 2013;25:973–981.

132. Niehaus TD, Kinison S, Okada S, Yeo YS, Bell SA, Cui P, Devarenne TP, Chappell J. Functional identification of triterpene methyltransferases from Botryococcus braunii Race B. J. Biol. Chem. 2012;287:8163–8173.

133. Hillen LW, Pollard G, Wake LV, White N. Hydrocracking of the oils of Botryococcus braunii to transport fuels. Biotechnol. Bioeng. 1982;24:193–205.

134. Pulis AP, Aggarwal VK. Synthesis of enantioenriched tertiary boronic esters from secondary allylic carbamates.Application to the synthesis of C30 botryococcene. J. Am. Chem. Soc. 2012;134:7570–7574.

135. Davies JJ, Krulle TM, Burton JW. Total synthesis of 7, 11-cyclobotryococca-5, 12, 26-triene using an oxidative radical cyclization as a key step. Org. Lett. 2010;120:2738–2741.

136. Hird NW, Lee TV, Leigh AJ, Maxwell JR, Peakman TM. The total synthesis of 10-(R,S)-C-30 botryococcene and botryococcane and a new synthesis of a general intermediate to the botryococcene family. Tetrahedron Lett. 1989;30:4867–4870.

137. White JD, Reddy GN, Spessard GO. Total synthesis of (-)-botryococcene. J. Am. Chem. Soc. 1988;110:1624–1626.

138. White JD, Reddy GN, Spessard GO. Total synthesis of (-)-C-34-botryococcene, the principal triterpenoid hydrocarbon of the freshwater alga Botryococcus braunii. J. Chem. Soc. Perk. T 1. 1993:759–767.

139. Dall'Osto L, Holt NE, Kaligotla S, Fuciman M, Cazzaniga S, Carbonera D, Frank HA, Alric J, Bassi R. Zeaxanthin protects plant photosynthesis by modulating chlorophyll triplet yield in specific light-harvesting antenna subunits. J. Biol. Chem. 2012;287:41820–41834.

140. Shegokar R, Mitri K. Carotenoid lutein: a promising candidate for pharmaceutical and nutraceutical applications. J. Diet. Suppl. 2012;9:183–210.

141. Yamaguchi M. Role of carotenoid beta-cryptoxanthin in bone homeostasis. J. Biomed. Sci. 2012;19

142. Di Mascio P, Kaiser S, Sies H. Lycopene as the most efficient biological carotenoid singlet oxygen quencher. Arch. Biochem. Biophys. 1989;274:532–538.

143. Shen RP, Jiang XY, Ye WD, Song XH, Liu L, Lao XJ, Wu CL. A novel and practical synthetic route for the total synthesis of lycopene. Tetrahedron. 2011;67:5610–5614.

144. Landrum JT, Bone RA. Lutein, zeaxanthin, and the macular pigment. Arch. Biochem. Biophys. 2001;385:28–40.

145. Khachik F, Chang AN. Total synthesis of (3R,3 ' R,6 ' R)-Lutein and its stereoisomers. J. Org. Chem. 2009;74:3875–3885.

PART IV

GENETIC ENGINEERING

CHAPTER 8

Metabolic Process Engineering for Biochemicals and Biofuels Production

SHANG-TIAN YANG AND XIAOGUANG LIU

8.1 INTRODUCTION

The trends towards using green chemical and energy are increasing due to the growing demand for non-fossil bio-products, the environmental concern for using fossil fuel, and the continuously increasing cost of crude oils. The total annual markets of biofuels and biochemicals are estimated to exceed $1 trillion [1]. Microbial fermentation has been widely used to produce organic biochemicals and biofuels, including citric acid, lactic acid, butyric acid, propionic acid, amino acids, ethanol, propanol, butanol, etc. The rapidly growing biotechnology market requires an efficient bioprocess platform, including both the production cell and the process, for biochemicals and biofuels production. Metabolic engineering (ME) is often used to develop high-producing cells needed for the process. However, ME requires genetically modifying the cell, which can be difficult to do or

Metabolic Process Engineering for Biochemicals and Biofuels Production. © *2014 Yang ST, et al.* J Microb Biochem Technol *6: e116. doi:10.4172/1948-5948.1000e116. Creative Commons Attribution License.*

to achieve the expected outcome, especially for less studied microorganisms. Metabolic process engineering (MPE) is a novel and advanced technology that alters or manipulates metabolic pathway to produce the interested metabolites by rationally controlling or manipulating bio-production process parameters. The goal of MPE is to achieve a high-productivity, high-quality, robust and scalable process through dynamic monitoring and investigating the interactions between cellular metabolism and process parameters. Different from the well-known traditional fermentation process development, MPE targets to engineer the bio-production process by controlling the cell physiology and metabolic responses to changes in fermentation process parameters and incorporating the interplay between cell and process into the rational process design. In this article, we focus on the application of MPE to improve biochemicals and biofuels production via precise bioreactor controllers, in situ sensors, and omics technologies.

Fermentation can be disturbed by slight changes in some process parameters, which leads to variable product quality. The major process parameters in fermentation include bioreactor operation parameters and metabolic process parameters. The bioreactor operation parameters (e.g., agitation rate, temperature, pH and DO) can be controlled by precise bioreactor controllers and in situ probes, which have been well evaluated and developed in process development in the biotechnology industry. For example, a fuzzy-PI controller has been developed to maintain a precise temperature by controlling temperature variation within a narrow range in large scale ethanol production [2]. The mathematic modelling has been successfully developed to assess the dynamic behaviour of bio-butanol fermentation consisting of various interconnected units such as fermenter, cell retention system, and vacuum vessel [3]. However, it is hard to directly regulate or manipulate metabolic process parameters (e.g., basal medium, substrate, feed rate and feed formulation) in fed-batch fermentation due to complicated and dynamically variable metabolic activities of microorganism.

Fed-batch fermentation has been widely used in biochemical, biofuel and food industries. The optimal nutrient feeding strategy can main certain cell growth to support bio-production, avoid nutrient depletion to achieve high volumetric productivity, and minimize the accumulation of by-products. One challenge to optimize feeding strategy (e.g., feeding rate and feed formulation) is how to collect and analyse dynamic metabolic parameters, including biomass, extracellular metabolites (substrate, product, by-product, and other metabolite) and intracellular metabolites. Although HPLC, GC, MS and other analytical technologies have been applied to analyse substrates and products offline, the profiling of intracellular and extracellular metabolites (e.g. the interested intermediates and end metabolite in key metabolic pathway), varies in the course of time. Robust, fast response, precise and in situ probes can partially solve this issue by providing online sample analysis. Some sensors, including biomass probe, dissolved oxygen probe, extracellular oxidoreduction potential probe, and gas monitors, are used in fermentation to monitor cell growth, assess aerobic metabolism, estimate NAD+/NADH+ ratio, and measure gases (e.g., CO_2, CH_4 and H_2), respectively [4]. In addition, biochemical analysers connected to auto-samplers are used for online monitoring of multiple metabolites. Different from discrete sensors, the MS based chemical multisensory systems, named as electronic noses and electronic tongues [5], have been used in both qualitative recognition of multi-component media and quantitative analysis of component concentrations in wine production. With dynamic metabolite data collected using in situ probes and auto-samplers, mathematic modelling can be applied to achieve precise nutrients feed and harvest time in fedbatch fermentation. For example, the maximum or dynamic substrate feed rate in aerobic fermentation has been determined by developing a feeding model that correlates substrate mass transfer and substrate uptake to volumetric oxygen transfer rate [6].

High-productivity fermentation processes have been developed using traditional bioreactor controller and metabolic process development tools. However, the rational design of a metabolically engineered fermentation process to achieve high-productivity, highquality, and high-robustness is far behind in the biotechnology industry. This is caused by the lack of a fundamental understanding of the interaction between cellular activities and fermentation environment. The recent advances in omics technologies enable the fermentation process profiling, and thereby provides an in-depth understanding of genome background, global protein profiling, and metabolite map of bio-production. Omics studies usually refer to genomics, transcriptomics, proteomics, metabolomics and others. 1) Genomics is the comprehensive and complete analysis of genome using new-generation

DNA sequencers such as Illumina Hi Seq 2000 or Life Tech SOLiD; 2) Transcriptomics is a functional genomics analysis by qualifying and quantitating messenger RNA using next-generation sequencing technologies; 3) Proteomics is to quantitate the expression of intracellular proteins under defined culture conditions using SELDITOF- MS, UPLC-MS/MS and MALDI-TOF-MS; and 4) Metabolomics is to identify and quantify a large number of cellular metabolites using LC-MS.

Omics have been recently used in the biotechnology industry to develop fundamental understanding of the phenotype in biobutanol and biochemical production. For example, transcriptomics has been used to analyse the response of *Clostridium acetobutylicum* ATCC 824 to butanol stress, which generated a new medium formulation to maintain high cell growth and butanol production [7]. Another genome-wide transcriptional analysis with the next-generation sequencing technology has been performed to investigate the effect of butyrate supplement on butanol metabolic switch in *C. beijerinckii* NCIMB 8052 [8]. With the access and integration with genomics database and transcriptomics knowledge, it is feasible to identify metabolites, establish metabolic reactions, and reconstruct metabolic networks via metabolomics. The core metabolites responsible for carbon, energy and redox balance, amino acids, end product inhibition and cell growth under defined culture conditions or production processes can also be distinguished.

Metabolomics is a powerful approach in MPE because it is capable of finding the regulatory mechanism of metabolic flux balance or regulation. Metabolic flux reveals the overall outcome of various cellular components, such as genes, transcripts, proteins, and metabolites, and interplayed factors, such as gene regulation, proteinprotein interaction, and metabolic network. Therefore, the metabolic flux analysis facilitated with metabolomics approach is the key to MPE. The increasing metabolic coverage and analytical resolution in metabolomics provides the direct evaluation of pathway intermediates. Multiple software tools (e.g., Open Flux and Fiat-Flux) are available and allow for user-friendly metabolic flux calculation by integrating experimental metabolomics data [9,10]. The metabolic network can be constructed using statistical analysis such as unsupervised learning, correlation network analysis, pattern recognition, principle component analysis, or dynamics control theory [11].

-

With the rapid advancement of systems biology, a large amount of metabolomics data has been accumulated and some well-known public metabolic pathway databases have been created, such as, MetaCyc, Kyoto Encyclopedia of Genes and Genomes (KEGG), Pathway Interaction Database (PID), Reactome and WikiPathway [12]. Some de novo models have been developed to facilitate data interpretation, but they rely on the literature mining and manual processing, so it is still challenging to extract key information from the big data [13]. To solve this issue, Buchel, et al. have established Path2Models database by including kinetic, logical, rule-based, multi-agent, constraint-based and statistical models [14]. The advantage of Path2Models database is that it can automatically generate mathematical model from pathway data sources, such as KEGG, Bio Carta, Meta Cyc and SABIO-RK. Various types of models have been developed based on the Path 2 Models and shared through Bio Modes Database and the Cell ML repository [11]. In addition to these databases and models, computational systems biology modelling will be a good strategy to perform functional analysis and infer cellular network, which integrates various statistical frameworks and mathematical formulas [13].

With the continuing market growth for microorganism-based biochemicals and biofuels, it is of great interest for the biotechnology industry to rationally design effective bioprocesses. Rational design requires the accurate prediction of cell responses to changes in fermentation conditions. The rational process design empowered by omics technologies, especially transcriptomics and metabolomics, allows for investigating gene expression, developing metabolite profiling, distinguishing metabolic regulators, and identifying critical process parameters. Therefore, the integration of rational design with omics technologies in MPE can contribute to the development of metabolically engineered processes for industrial production of biochemicals and biofuels with high productivity and high product quality.

In summary, MPE is a powerful technology that integrates the well-developed process control techniques, such as precise bioreactor controllers and in situ sensors, and advanced omics technologies. MPE enables the rational design of a bio-production process, and thus can lead to a highly efficient fermentation process for biochemicals and biofuels production. MPE not only can contribute to the enhanced production of metabolites in fermentation but also can provide an indepth understanding of interplays between cells and the fermentation process. Current metabolic engineering approaches require genetically modifying the cell, which can be difficult to do for less studied microorganisms. MPE is easier to implement than metabolic engineering and should have broad applications in biotechnology for the production of chemicals, fuels, and pharmaceuticals.

REFERENCES

1. Silicon Valley Bank (2012) Silicon Valley Bank Cleantech Practice. The Advanced Biofuel and Biochemical Overview.
2. Fonseca RR, Schmitz JE, Fileti AM, da Silva FV (2013) A fuzzy-split range control system applied to a fermentation process. BioresourTechnol 142:475-482.
3. Mariano AP, Costa CB, Maciel MR, MaugeriFilho F, Atala DI, et al. (2010) Dynamics and control strategies for a butanol fermentation process. ApplBiochemBiotechnol 160:2424-2448.
4. Du C, Yan H, Zhang Y, Li Y, Cao Z (2006) Use of oxidoreduction potential as an indicator to regulate 1,3-propanediol fermentation by Klebsiellapneumoniae. ApplMicrobiolBiotechnol 69:554-563.
5. Peris M, Escuder-Gilabert L (2013) On-line monitoring of food fermentation processes using electronic noses and electronic tongues: a review. Anal ChimActa 804:29-36.
6. Johnsson O, Andersson J, Liden G, Johnsson C, Hagglund T (2013) Feed rate control in fed-batch fermentations based on frequency content analysis. BiotechnolProg 29:817-824.
7. Heluane H, Evans MR, Dagher SF, Bruno-Barcena JM (2011) Meta-analysis and functional validation of nutritional requirements of solventogenic Clostridia growing under butanol stress conditions and coutilization of D-glucose and D-xylose. Appl Environ Microbiol 77:4473-4485.
8. Wang Y, Li X, Blaschek HP (2013) Effects of supplementary butyrate on butanol production and the metabolic switch in Clostridium beijerinckiiNCIMB 8052: genome-wide transcriptional analysis with RNA-Seq. Biotechnol Biofuels 6(1):138.
9. Quek LE, Wittmann C, Nielsen LK, Kromer JO (2009) OpenFLUX: efficient modelling software for 13C-based metabolic flux analysis. Microb Cell Fact 8:25.
10. Zamboni N, Fischer E, Sauer U (2005) FiatFlux--a software for metabolic flux analysis from 13C-glucose experiments. BMC bioinformatics 6:209.
11. Kohlstedt M, Becker J, Wittmann C (2010) Metabolic fluxes and beyond-systems biology understanding and engineering of microbial metabolism. ApplMicrobiolBiotechnol 88:1065-1075
12. Buchel F, Rodriguez N, Swainston N, Wrzodek C, Czauderna T, et al. (2013) Path-2Models: large-scale generation of computational models from biochemical pathway maps. BMC SystBiol 7:116.

13. Hyduke DR, Lewis NE, Palsson BO (2013) Analysis of omics data with genome-scale models of metabolism. MolBiosyst 9:167-174.
14. http://www.ebi.ac.uk/biomodels-main/path2models

CHAPTER 9

Enhanced Genetic Tools for Engineering Multigene Traits into Green Algae

BETH A. RASALA, SYH-SHIUAN CHAO, MATTHEW PIER, DANIEL J. BARRERA, AND STEPHEN P. MAYFIELD

9.1 INTRODUCTION

Microalgae have recently attracted attention as potential low-cost platform for the production of a broad range of commercial products including biofuels, nutraceuticals, therapeutics, industrial chemicals and animal feeds [1]–[11]; and genome engineering will enable and enhance algae-produced bio-products [1], [5], [6], [12]–[19]. However, while much has been written about the potential of transgenic microalgae, little of that potential has yet to be commercialized. A major obstacle to generating useful transgenic algae strains has been the lack of molecular tools and overall poor expression of heterologous genes from the nuclear genome of many microalgae species, at least partially due to rapid gene silencing [20]–[23]. For example, a set of validated vectors for targeting transgene products to specific subcellular locations do not exist, nor does the vector to allow the expression of multiple nuclear-encoded genes within a single cell.

Enhanced Genetic Tools for Engineering Multigene Traits into Green Algae. © 2014 Rasala et al. PLoS one 9(4): e94028. doi:10.1371/journal.pone.0094028. Creative Commons Attribution License.

Previously, we described a nuclear expression strategy that overcomes transgene silencing by using the foot-and-mouth-disease-virus (FMDV) 2A "self-cleaving" peptide to transcriptionally fuse transgene expression to the antibiotic resistance gene ble in the green microalga *Chlamydomonas reinhardtii* [23], [24]. It is believed that the FMDV 2A sequence "self-cleaves" through ribosome-skipping during translation rather than a proteolytic reaction, and has been termed CHYSEL (cis-acting hydrolase element) [25], [26]. This strategy allowed for the selection of transgenic lines that efficiently express the transgene-of-interest, and this robust expression remains for many generations. We demonstrated the utility of our pBle-2A vector with the expression and secretion of the valuable industrial enzyme, xylanase [23]. Furthermore, this expression strategy enabled, for the first time, the robust expression of six fluorescent proteins (FPs) in the cytosol of green microalgae [24]. FPs have become essential research tools that have revolutionized many fields of biology

Here we report the construction and validation of a set of transformation vectors that enable protein targeting to distinct subcellular locations, and present two complementary methods for multigene engineering in the eukaryotic green microalga *C. reinhardtii*.

9.2 RESULTS

Here we describe vectors that enable protein targeting to four important organelles: the nucleus, mitochondria, endoplasmic reticulum (ER), and chloroplast (Table 1). The nucleus houses the majority of the cell's genetic material, and therefore is critical for the regulation of most gene expression. To generate a nucleus-targeting vector, a tandem copy of the nuclear localization signal (NLS) from simian virus 40 (SV40) [27] was fused to the C-terminus of mCerulean, and transcriptionally linked to ble-2A (Figure 1A). Cells transformed with mCerulean-2xNLS displayed fluorescence signals that were concentrated in the nucleus (Figure 1E-G, Figure S1A), as confirmed by co-staining fixed mCerulean fluorescent cells with the nuclear DNA-stain Hoechst (Figure S1B).

Mitochondria function in respiration, producing ATP via oxidative respiration, and therefore play an essential role in cell metabolism.

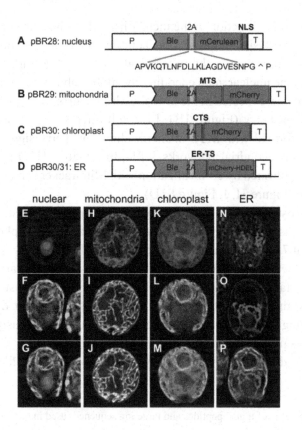

FIGURE 1: *Chlamydomonas* transformation vectors for protein targeting to specific subcellular locations. A–D. Schematic representation of *Chlamydomonas* targeting vectors. All transformation vectors contain the hsp70/rbcs2 promoter (P), the ble gene that confers resistance to zeocin, the 2A self-cleaving sequence from foot-and-mouth-disease virus, and the rbcs2 terminator (T). The site of cleavage is indicated with an arrowhead. A. pBR28, mCerulean is targeted to the nucleus by a C-terminal fusion to 2xSV40 NLS. B. pBR29, mCherry is targeted to the mitochondria by an N-terminal fusion to the mitochondrial transit sequence (MTS) of mitochondrial atpA. C. pBR32, mCherry is targeted to the chloroplast using the chloroplast transit sequence (CTS) from psaD. D. pBR30/31, mCherry is targeted to the ER using the ER-transit sequence (ER-TS) from either BiP1 or ars1. The ER retention sequence H-D-E-L is fused to the C-terminus of mCherry. E–P. Microscopy images of cells transformed with pBR28 (E–G), pBR29 (H–J), pBR32 (K–M), and pBR30 (N–P). Top row are live cell images of the fluorescent proteins targeted to the nucleus (E), mitochondria (H), chloroplast (K) or ER (N, O). (I) The cell is co-stained with the mitochondrial dye Mitotracker. (N) Cross section through the top of a cell expressing mCherry in the ER allows for the visualization of the cortical ER network. (O) Cross section through the middle of the same cell as in (N). The chloroplast membranes are visualized in (F), (L) and (P). Merged images are shown in the bottom row.

Mitochondria also function in the metabolism of amino acids, lipids, iron, calcium homeostasis, apoptosis and cell signaling. To generate a mito-chondria-targeting vector, the N-terminal mitochondria transit sequence (MTS) from the nuclear gene encoding the alpha subunit of the mitochon-drial ATP synthase located in the mitochondrial matrix, was fused between ble-2A and mCherry (Figure 1B). Live cell microscopy of independent clones transformed with MTS-mCherry shows mCherry signal localized to tubular mitochondrial networks [28]–[30] (Figure 1H, Figure S1D), which was confirmed by co-localization with Mitotracker, a mitochondria-specific dye (Figure 1I, J, Figure S1D).

The most studied *C. reinhardtii* organelle is the chloroplast, the site of photosynthesis. *C. reinhardtii* has a single cup-shaped chloroplast that oc-cupies about 75% of the volume of the cell. The chloroplast is also the site of multiple metabolic reactions, including the biosynthesis of amino acids, isoprenoids, fatty acids, and starch [31]. The chloroplast transit sequence (CTS) from the photosystem I protein psaD was chosen for the chloro-plast-targeting vector (Figure 1C), as it has been used previously to target heterologous proteins to the chloroplast in *C. reinhardtii* [32]. Cells trans-formed with the ble2A-CTS-mCherry vector displayed red fluorescence

TABLE 1: Summary of transit peptides and targeting sequences used in this study.

Vector	Location	Transit sequence	Size	Function	Reference
pBR28	Nuclear	2x SV40 NLS	20 aa	Tandem copy of a nuclear localization sequence from SV40	commonly used in mammalian vectors
pBR29	Mitochondria	atpA	45aa	Alpha subunit of mitochondrial ATP synthase.	this report
pBR30	ER	BIP1	31aa	Chaperone, Hsp70 superfamily.	this report
pBR31	ER	ARS1	30aa	Periplasmic protein involved in mineral-ization of sulfate by hydrolyzing sulfate esters.	Rasala et al., 2012
pBR32	Chloroplast	PSAD	35aa	Protein of Photosystem I.	Fischer and Rochaix, 2001

signals that properly localized to the chloroplast (Figure 1K, Figure S1E) and partially overlapped with chloroplast auto-fluorescence derived from chlorophyll and other pigments localized in chloroplast photosynthetic membranes (Figure 1L, M)

The endoplasmic reticulum (ER) forms an extensive interconnected network of tubules and flattened stacks located throughout the cytoplasm and is continuous with the nuclear envelope [33]. The ER has multiple functions, including translocation and modification of proteins destined for secretion. It also functions in lipid metabolism, carbohydrate metabolism, and detoxification. Two ER-targeting vectors were created by fusing the ER signal sequence of the *C. reinhardtii* genes ars1 or bip1 between ble-2A and mCherry; and the ER retention signal His-Asp-Glu-Leu (HDEL) [34] was fused to the C-terminus of mCherry (Figure 1D). ars1 encodes for a secreted arylsulfatase [35], and its signal peptide has been shown to target heterologous xylanase 1 for secretion [23]. BiP1 is an ER-localized chaperone of the HSP70 superfamily [36]. Using live cell microscopy, cells transformed with either ER-targeting vector displayed mCherry localization to reticular, net-like structures under the plasma membrane, which are reminiscent of cortical ER that has been characterized in other eukaryotes (Figure 1N). Z planes focused through the middle of cells demonstrate that mCherry localizes to a structure that is continuous with the nuclear envelope (Figure 1O, P, Figure S1C). While we were unable to identify any ER-specific dyes that function in *Chlamydomonas*, we are confident that both ars1-mCherry-HDEL and BiP1-mCherry-HDEL successfully targets the FP to the ER, based on the resultant distinct and characteristic localization pattern.

To verify the fluorescence live cell microscopy data, SDS-PAGE immunoblotting was performed on lysates from individual transformants expressing the targeting vectors described above. Immunoblots demonstrate that the targeted fluorescent proteins accumulate to detectible levels, are correctly processed from ble-2A, and display the predicted mobility for the respective mature protein (Figure S2).

Several successful strategies for the coordinated expression of multiple genes have been described in other eukaryotes. These include the use of FMDV-2A and 2A-like peptides to ensure transcriptional co-expression of multiple proteins encoded in a single open reading frame (ORF) with co-translational "cleavage" into distinct peptides [26]. However, with the exception of a gene-of-interest and an antibiotic resistance marker, coordinated multi-gene expression has yet to be achieved in green microalgae. We modified our ble2A expression strategy to include a second 2A peptide from equine rhinitis A virus (E2A) [37] followed by a third protein coding sequence (Figure 2A). Transgenic algae expressing the Ble•E2A−mCerulean•2xNLS•F2A−BiP•mCherry•HDELORF were recovered which had properly integrated the multi-cistron transgene cassette and accumulated mCerulean in the nucleus and mCherry in the ER (Figure 2B, C), both at high levels of expression.

A potential disadvantage of the double 2A vector is that the 2A C-terminal fusion to the middle protein of the poly-cistron may disrupt its function and/or localization. Thus, we developed a gene-stacking strategy to generate transgenic algae that express up to four targeted proteins by harnessing the power of genetic breeding. *C. reinhardtii* is a haploid organism that normally divides vegetatively. However, under certain conditions, a mating-type plus (mt+) gamete will mate with a mating-type minus (mt-) gamete to form a diploid zygospore that then undergoes meiosis to yield four haploid progeny. During this mating, genes integrated into separate chromosomes can individually assort resulting in progeny with genes from either parent. To test whether we could cross two transgenic lines and obtain a single progeny that contained both transgenes that were still expressed at desirable levels, we mated an mt+ strain that expressed mCherry targeted to the ER (Figure 3A) to an mt- strain that expressed mCerulean targeted to the nucleus (Figure 3B), both as ble2A fusions. The progeny were FACS sorted for cells that expressed both mCerulean and mCherry. The presence and expression of both inherited transgenes was verified by PCR analysis and fluorescence microplate reader analysis (Figure 3H and data not shown). Live cell microscopy confirmed that both engineered genes from the transgenic parents, nuclear-localized mCerulean and ER-localized mCherry, were inherited in selected progeny (Figure 3C). This process was repeated, mating the two-colored algae to an additional transgenic line expressing mitochondria-targeted Venus (Figure 3D) to obtain progeny that robustly expressed three engineered transgenes within a single cell (Figure 3E). After a third round of mating between the 3-colored strain with a strain that stably express alpha-tubulin fused

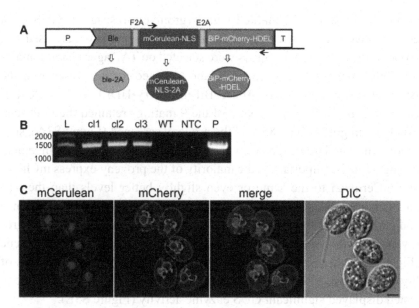

FIGURE 2: Gene stacking using a multi-cistronic transformation vector. A. A schematic representation of the *Chlamydomonas* multi-cistronic expression vector. The expression of the cassette is under the control of the hsp70/rbcs2 promoter (P). Ble confers zeocin-resistance. mCerulean is targeted to the nucleus with the SV40-NLS. mCherry is targeted to the ER using the BiP ER-TS and the HDEL retention sequence. Two 2A self-cleaving sequences are fused between the three cistrons: F2A, from FMDV1; E2A, from equine rhinitis A virus. Black arrows represent the location of the oligonucleotides used in (B). Following co-translational processing of the 2A peptides, three distinct proteins are expressed (ovals). B. PCR analysis of the multi-cistron cassette genome integration. Transformants were screened by PCR to identify individual clones that correctly integrated the multi-cistronic transformation vector, using the oligonucleotides indicated by the arrows in (A). Three independent clones (cl) are shown. L, ladder; WT, wildtype cc1690; NTC, no template control; P, plasmid. C. Live cell fluorescence microscopy of a clone expressing the multi-cistronic vector. mCerulean-NLS (blue) localizes to the nucleus while BiP-mCherry-HDEL (red) is targeted to the ER. Scale bar, 5 μm.

to mTagBFP (Figure 3F), we obtained progeny that expressed four different FPs, all properly localized to four distinct subcellular locations: the nucleus, ER, mitochondria, and flagella (Figure 3G).

To determine whether the two gene stacking approaches described above could be used in combination, we mated three independent clones that stably express proteins from the multi-cistron double-2A vector, to three independent clones of the opposite mating type that express β-glucuronidase (GUS) marked with hygromycin-resistance. GUS is an enzyme involved in the catalysis of carbohydrates and is widely used as a reporter. Progeny from the cross were selected on TAP agar plates containing both zeocin and hygromycin B, and screened for the presence of the Ble•E2A−mCerulean•2xNLS•F2A−BiP•mCherry•HDELORF by PCR as described above. Progeny from 8 of the 9 matings retained the multi-cistron ORF in greater than 85% of the progeny screened (Figure S3A), suggesting that the large expression cassette does not undergo rearrangement during meiosis. Importantly, the majority of the progeny express mCherry and mCerulean to the same or even slightly better levels than the parents (Figure S3C, D). The progeny were also screened for the presence and expression of the GUS gene. As expected, the majority of the hygromycin-resistant progeny also retained the genetically linked GUS gene (Figure S3B). Interesting, while the parents displayed poor expression of β-glucuronidase likely due to transgene silencing, most of the progeny tested displayed significant GUS enzyme activity (Figure S3E).

9.3 DISCUSSION

The use of transgenic microalgae for the production of bioproducts has enormous economic and biotechnology promise, because algal production combines the simplicity and speed of haploid, single-cell genetics in an organism with elaborate biosynthetic potential, and with the associated economic benefit of using photosynthesis to drive product formation. Here we describe key genetic tools that will enable complex genetic and metabolic engineering in green microalgae: the ability to target gene products to specific subcellular locations, and vectors and well-characterized protocols that enable multi-gene stacking within a single transgenic cell. We have generated a class of nuclear transformation vectors that efficiently and specifically target transgene products to the nucleus, mitochondria, ER and chloroplast (Figure 1). Furthermore, the transit sequences can be used in combination with multiple 2A self-cleaving sequences to generate multi-cistron vectors that enable robust and coordinated expression of multiple recombinant proteins from a single transcript (Figure 2). Finally, we describe methods to stack up to four transgenes within a single cell

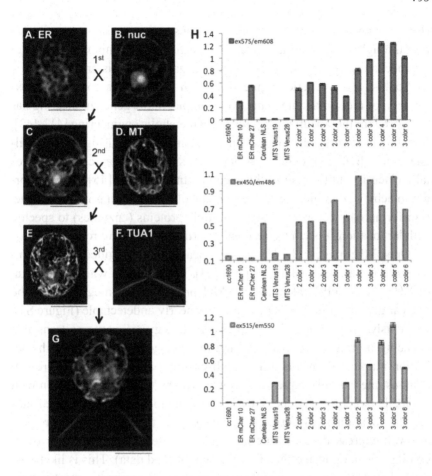

FIGURE 3: Gene stacking through mating. An mt+ strain transformed with pBR30 (A) was crossed with an mt- strain transformed with pBR28 (B). Progeny that expressed mCherry in the ER and mCerulean in the nucleus were obtained (C). Cell lines expressing both ER-mCherry and nuclear mCerulean were crossed with cells transformed with mitochondria-targeted Venus (D), to obtain progeny that stably expressed three distinct FPs in three sub-cellular locations (E). These cell lines were crossed to transgenic cells that expressed α-tubulin (TUA1) fused to mTagBFP (F), to obtain progeny that expressed four FPs in four distinct subcellular locations (G). H. Fluorescence plate reader assays of the parents and progeny indicate that FP expression remains stable following matings. Cell lines were assayed for mCherry expression (ex575/em608), mCerulean expression (ex450/em486), and Venus expression (ex515/em550). ER-mCherry parents (ER-mCher 10 and 27), nuclear mCerulean parent (Cerulean NLS) and mitochondrial Venus parents (MTS Venus 19 and 28) are shown along with WT cc1690. 2 color cell lines 1–4 express ER-mCherry and Cerulean-NLS. 3 color cell lines 1–6 express ER-mCherry, nuclear mCerulean and mitochondrial Venus. Scale bars, 5 μm.

(Figure 3 and Figure S3). Importantly, transgene expression remains robust throughout the mating process, suggesting that the microalgal silencing mechanism(s) are not activated during gametogenesis or meiosis.

Sophisticated genetic engineering often requires the coordinated expression of more than one gene. For example, multi-gene engineering has been used for therapeutics [38]–[40], and metabolic engineering [41]–[43]. Metabolic networks are complex in all eukaryotic organisms including algae, and individual biochemical steps of a single pathway can sometimes take place in multiple subcellular compartments [44], [45]. Thus, in order to achieve complex genetic and metabolic engineering in microalgae, transformation vectors that target multiple proteins (enzymes) to specific cellular locations – such as the ones described above - are required.

One of the biggest challenges to nuclear genome engineering in *C. reinhardtii* is transgene silencing [20]–[23]. For example, when mCerulean or GUS are directly linked to the PAR1 promoter and integrated into the nuclear genome, the reporter proteins are nearly undetectable (Figure S4). Previously, we developed an expression strategy to overcome transgene silencing by transcriptionally linking the transgene-of-interest to the selection marker Ble through a 2A self-cleaving sequence [23]. Here, we demonstrate that this expression strategy can be used in combination with organelle targeting sequences to direct protein localization to desired subcellular locations. Furthermore, our data show that the targeted proteins are well-expressed. Ble is the most effective selection marker for overcoming silencing tested thus far (our unpublished data). This is likely because Ble functions by sequestration rather than enzymatic inactivation, binding to zeocin in a 1-to-1 ratio [46]. Thus high levels of Ble expression are required to survive zeocin selection. However, the ability to use only one selection marker limits the utility of the ble-2A expression strategy. The double 2A multi-cistron vector was developed to overcome this limitation. Indeed, the multi-cistron vector was used to co-express two reporter proteins that were directed to two distinct subcellular locations. We further demonstrate that the double 2A vector is stable; the similar 2A sequences do not recombine during vegetative cell division or meiosis to loop out the middle coding sequence.

A second gene-stacking strategy investigated was gene-stacking though mating. Two strains of opposite mating types engineered to express ble-2A-ER-mCherry or ble-2A-mCerulean-NLS were mated, germinated on TAP/zeocin plates and then FACS sorted for mCherry and mCerulean. Even though progeny were selected on zeocin, either ble2A construct could provide antibiotic resistance. Thus, there was only selection pressure for the expression of one – and not both – of the ble2A constructs. Notably, however, we were able to recover strains that robustly expressed both mCherry and mCerulean. This result was repeated with three and then four transgenes. Indeed, even when we were unable to distinguish mTagBFP from mCerulean by FACS and therefore progeny could not be enriched by flow cytometry, we were still able to easily recover strains in which mCerulean-NLS and mTagBFP-TUA were well expressed. These data suggest that the robust silencing mechanisms that are well-described but poorly understood may not affect transgenes once they have escaped the initial mechanism of silencing upon transformation and integration.

Indeed, our data indicate that transgene silencing may even be lessened following mating and meiosis. Two independent transgenic strains expressing a silenced GUS gene were mated and GUS-positive progeny were assayed for GUS expression (Figure S3). Most of the progeny from both crosses displayed significantly more GUS activity than the parent strains. We are currently investigating the molecular mechanism behind these notable results.

Microalgae are poised to revolutionize many industries including energy, nutrition, health, and specialty chemicals. The molecular genetic tools and methods described here for multi-gene engineering and protein targeting will significantly advance the current state of microalgae genetic, metabolic, and pathway engineering, and therefore impact the development of transgenic algae as a biotechnology platform.

9.4 MATERIAL AND METHODS

9.4.1 ALGAL STRAINS, TRANSFORMATIONS AND GROWTH CONDITIONS

The *C. reinhardtii* strains used in this study were cc1690 (mt+) and cc1691 (mt-, *Chlamydomonas* Resource Center). Cells were transformed by electroporation as described previously [23]. Transformants were selected on TAP (Tris–acetate–phosphate) agar plates supplemented with 2.5–10 µg/ml zeocin. Transformants were screened by PCR to identify gene positive transformants as described previously [23].

9.4.2 PLASMID CONSTRUCTION

mCherry, Venus, mCerulean, and mTagBFP were codon-optimized for expression from the nuclear genome of *C. reinhardtii*, as previously described [24]. The organelle transit sequences from bip1, psaD, and atpA were PCR-amplified from genomic DNA isolated from cc1690, using the oligonucleotides described in Table S1, and fused between ble2A and mCherry in the pBR9 vector [24] using the GeneArt Seamless Cloning Kit (Life Technologies, Carlsbad, CA). The 2x SV40 NLS was codon-optimized for *C. reinhardtii* nuclear expression, synthesized as sense and antisense single stranded oligonucleotides, annealed, and cloned into pBR25 [24] that had been digested with BamHI and EcoRI. His-Asp-Glu-Lys (HDEL) was fused to the end of mCherry by PCR using a reverse oligonucleotide encoding for the ER retention sequence. mCherry-HDEL was cloned behind Ble-2A-BiP or Ble-2A-ARS1 [23] by restriction digest and ligation using the enzymes XhoI and BamHI. To generate pBR26 double-2A vector, the 2A sequence from equine rhinitis A virus (E2A) [37] was first codon-optimized, synthesized, and tested for self-cleavage in *C. reinhardtii* (our unpublished data). pBR30 was linearized by PCR and mCerulean-NLS and E2A were fused between Ble2A and BiP-mCherry-HDEL using the GeneArt Seamless Plus Cloning Kit (Life Technologies). GUS was codon-optimized for expression from the nuclear genome, synthesized and cloned into the pBR2 hygromycin resistance expression cassette [23] by restriction digest with NdeI and BamHI.

9.4.3 FLUORESCENCE MICROSCOPY

Representative clones were grown in TAP media without antibiotics to late log phase on a rotary shaker. Mitotracker Green FM (Life Technologies) was used to stain the mitochondria of live cells as per the manufacturer's instructions. Live cells were plated on TAP/1% agar pads prior to image acquisition. Images were captured on a Delta Vision (Applied Precision Inc., Issaquah, WA) optical sectioning microscope system composed of an Olympus IX71 inverted microscope (Center Valley, PA) equipped with an Olympus UPlanSApo 100×/1.40 objective and a CoolSNAP HQ2/ICX285 camera (Photometrics, Tucson, AZ). The following filters were used: mTagpBFP, excitation 360/40 nm, emission 457/50 nm; mCerulean, excitation 436/10 nm, emission 470/30 nm; Venus, excitation 470/40 nm, emission 515/30 nm; mCherry, excitation 558/28 nm, emission 617/73 nm; and Mitotracker Green FM, excitation 470/40 nm, emission 515/30 nm. Image acquisition and deconvolution were performed using Resolve3D SoftWoRx-Acquire (Version 5.5.1, Applied Precision Inc). Brightness and contrast were adjusted using Adobe Photoshop CS3 or ImageJ software. The images in Figure S1 were adjusted identically.

9.4.4 FLUORESCENCE MICROPLATE READER ASSAY

Cells were grown in TAP media without antibiotics until late log phase. 100 μls of cells were transferred, in triplicate, to wells of a black 96 well plate (Corning Costar, Tewksbury MA), and fluorescence was read using a Tecan plate reader (Tecan Infinite M200 PRO, Männedorf, Switzerland). Fluorescence readings with the indicated excitation/emission filters were acquired using a calculated optimal gain, which was determined prior to each reading. TAP media was used to blank the readings. Fluorescence signals were normalized by chlorophyll fluorescence (excitation 440/9 nm, emission 680/20 nm).

9.4.5 CHLAMYDOMONAS MATINGS

Matings were performed using the following protocol: gametes were generated by incubating mt+ and mt- cells overnight in nitrogen-free liquid TAP. Mt+ gametes were mixed with mt- gametes for 2–4 hours, and the mating reactions were plated to TAP/3% agar plates and incubated in the dark for 5–7 days. Unmated cells were scraped off to the side using a sterile razor blade and the plates were subjected to chloroform treatment to kill any remaining unmated cells. For the matings described in Figure 3, spores were collected using an inoculating loop and struck onto TAP agar plates supplemented with 10 μg/ml of zeocin and incubated in the light until colonies appeared. For the matings described in Figure S3, spores were inoculated into 50 ml TAP and grown in light until late log phase. Cells were then plated to TAP agar plates containing 10 μg/ml of zeocin and 15 μg/ml of hygromycin B.

9.4.6 FLUORESCENCE ACTIVATED CELL SORTING

Progeny from matings were inoculated into TAP liquid cultures without antibiotics and grown to late log phase, diluted back 1:10, and grown for another 12–24 hours. The cultures were then sorted for expression of the appropriate fluorescence proteins on a BD Influx cell sorter (BD Biosciences, Vannas, Sweden), gating for mCherry+ and mCerulean+ for the 2-color strain; and mCherry+, mCerulean+ and Venus+ for the 3- and 4-color strain. Sorted progeny were verified by PCR analysis to confirm the presence of the targeted-FP genes, and by plate reader assay prior to microscopy. mCherry-expressing cells were identified using a 532 nm laser and a 585/40 nm filter to detect red fluorescence. mCerulean-expressing cells were identified using a 457 nm laser and a 480/40 nm filter to detect cyan fluorescence. Venus-expressing cells were identified using a 488 nm laser for excitation and a 530/40 nm filter. mTagBFP fluorescence could not be distinguished from mCerulean fluorescence in the 4-color strain, rather cells were sorted for cyan/blue, yellow, and red fluorescence and then verified by PCR analysis and plate reader assay for the presence of all four desired FP genes.

9.4.7 GUS ACTIVITY ASSAY

Cell cultures were grown in liquid TAP media without antibiotics to late log phase in 12-well plates. 100 μls of cells were incubated with 25 μls of 1 mg/ml 4-methylumbelliferyl-beta-D-glucuronide (Sigma, St. Louis, MO) in a 96-well black microplate for 20 minutes. GUS activity was determined using a fluorometric assay, by measuring the accumulation of the fluorescent product (ex365/em455).

REFERENCES

1. Specht E, Miyake-Stoner S, Mayfield S (2010) Micro-algae come of age as a platform for recombinant protein production. Biotechnol Lett 32: 1373–1383 Available: http://www.pubmedcentral.nih.gov/articlerender.fcgi?artid=2941057&tool=pmcent rez&rendertype=abstract. Accessed 9 August 2013.

2. Jones CS, Mayfield SP (2012) Algae biofuels: versatility for the future of bioenergy. Curr Opin Biotechnol 23: 346–351 Available: http://www.ncbi.nlm.nih.gov/ pubmed/22104720. Accessed 14 August 2013.

3. Pulz O, Gross W (2004) Valuable products from biotechnology of microalgae. Appl Microbiol Biotechnol 65: 635–648 Available: http://www.ncbi.nlm.nih.gov/ pubmed/15300417. Accessed 17 August 2013.

4. Spolaore P, Joannis-Cassan C, Duran E, Isambert A (2006) Commercial applications of microalgae. J Biosci Bioeng 101: 87–96 Available: http://www.ncbi.nlm.nih.gov/ pubmed/16569602. Accessed 8 August 2013.

5. Raja R, Hemaiswarya S, Kumar NA, Sridhar S, Rengasamy R (2008) A perspective on the biotechnological potential of microalgae. Crit Rev Microbiol 34: 77–88 Available: http://www.ncbi.nlm.nih.gov/pubmed/18568862. Accessed 6 August 2013.

6. TL, Purton S, Becker DK, Collet C (2005) Microalgae as bioreactors. Plant Cell Rep 24: 629–641 Available: http://www.ncbi.nlm.nih.gov/pubmed/16136314. Accessed 9 August 2013.

7. Olaizola M (2003) Commercial development of microalgal biotechnology: from the test tube to the marketplace. 20: 459–466 Available: http://www.ncbi.nlm.nih.gov/ pubmed/12919832. Accessed 12 August 2013.

8. View Article PubMed/NCBI Google Scholar

9. 8. Christaki E, Florou-Paneri P, Bonos E (2011) Microalgae: a novel ingredient in nutrition. Int J Food Sci Nutr 62: 794–799 Available: http://www.ncbi.nlm.nih.gov/ pubmed/21574818. Accessed 27 August 2013.

10. Chisti Y (2008) Biodiesel from microalgae beats bioethanol. Trends Biotechnol 26: 126–131 Available: http://www.ncbi.nlm.nih.gov/pubmed/18221809. Accessed 7 August 2013.

11. Barnes D, Franklin S, Schultz J, Henry R, Brown E, et al. (2005) Contribution of 5'- and 3'-untranslated regions of plastid mRNAs to the expression of Chlamydomonas reinhardtii chloroplast genes. Mol Genet Genomics 274: 625–636 Available: http:// www.ncbi.nlm.nih.gov/pubmed/16231149. Accessed 5 August 2013.

12. Mayfield SP, Manuell AL, Chen S, Wu J, Tran M, et al. (2007) Chlamydomonas reinhardtii chloroplasts as protein factories. Curr Opin Biotechnol 18: 126–133 Available: http://www.ncbi.nlm.nih.gov/pubmed/17317144. Accessed 8 August 2013.

13. JN, Oyler GA, Wilkinson L, Betenbaugh MJ (2008) A green light for engineered algae: redirecting metabolism to fuel a biotechnology revolution. Curr Opin Biotechnol 19: 430–436. Available: http://www.ncbi.nlm.nih.gov/pubmed/18725295. Accessed 6 August 2013.

14. Radakovits R, Jinkerson RE, Darzins A, Posewitz MC (2010) Genetic engineering of algae for enhanced biofuel production. Eukaryot Cell 9: 486–501 Available: http://www.pubmedcentral.nih.gov/articlerender.fcgi?artid=2863401&tool=pmcentrez&rendertype=abstract. Accessed 9 August 2013.

15. Georgianna DR, Mayfield SP (2012) Exploiting diversity and synthetic biology for the production of algal biofuels. Nature 488: 329–335 Available: http://www.ncbi.nlm.nih.gov/pubmed/22895338. Accessed 8 August 2013.

16. Wijffels RH, Kruse O, Hellingwerf KJ (2013) Potential of industrial biotechnology with cyanobacteria and eukaryotic microalgae. Curr Opin Biotechnol 24: 405–413 Available: http://www.ncbi.nlm.nih.gov/pubmed/23647970. Accessed 7 August 2013.

17. Gimpel JA, Specht EA, Georgianna DR, Mayfield SP (2013) Advances in microalgae engineering and synthetic biology applications for biofuel production. Curr Opin Chem Biol 17: 489–495 Available: http://dx.doi.org/10.1016/j.cbpa.2013.03.038. Accessed 13 August 2013.

18. Beer LL, Boyd ES, Peters JW, Posewitz MC (2009) Engineering algae for biohydrogen and biofuel production. Curr Opin Biotechnol 20: 264–271 Available: http://www.ncbi.nlm.nih.gov/pubmed/19560336. Accessed 7 August 2013.

19. Gong Y, Hu H, Gao Y, Xu X, Gao H (2011) Microalgae as platforms for production of recombinant proteins and valuable compounds: progress and prospects. J Ind Microbiol Biotechnol 38: 1879–1890 Available: http://www.ncbi.nlm.nih.gov/pubmed/21882013. Accessed 6 August 2013.

20. Hannon M, Gimpel J, Tran M, Rasala B, Mayfield S (2010) Biofuels from algae: challenges and potential. Biofuels 1: 763–784 Available: http://www.pubmedcentral.nih.gov/articlerender.fcgi?artid=3152439&tool=pmcentrez&rendertype=abstract.

21. Cerutti H, Johnson AM, Gillham NW, Boynton JE (1997) Epigenetic silencing of a foreign gene in nuclear transformants of Chlamydomonas. Plant Cell 9: 925–945 Available: http://www.plantcell.org/content/9/6/925.abstract. Accessed 27 August 2013.

22. Fuhrmann M, Oertel W, Hegemann P (1999) A synthetic gene coding for the green fluorescent protein (GFP) is a versatile reporter in Chlamydomonas reinhardtii+. Plant J 19: 353–361 Available: http://doi.wiley.com/10.1046/j.1365-313X.1999.00526.x. Accessed 27 August 2013.

23. Neupert J, Karcher D, Bock R (2009) Generation of Chlamydomonas strains that efficiently express nuclear transgenes. Plant J 57: 1140–1150 Available: http://www.ncbi.nlm.nih.gov/pubmed/19036032. Accessed 21 August 2013.

24. Rasala BA, Lee PA, Shen Z, Briggs SP, Mendez M, et al. (2012) Robust expression and secretion of Xylanase1 in Chlamydomonas reinhardtii by fusion to a selection gene and processing with the FMDV 2A peptide. PLoS One 7: e43349 Available: http://www.pubmedcentral.nih.gov/articlerender.fcgi?artid=3427385&tool=pmcentrez&rendertype=abstract. Accessed 30 July 2013.

25. Rasala BA, Barrera DJ, Ng J, Plucinak TM, Rosenberg JN, et al. (2013) Expanding the spectral palette of fluorescent proteins for the green microalga Chlamydomonas reinhardtii. Plant J 74: 545–556 Available: http://www.ncbi.nlm.nih.gov/pubmed/23521393. Accessed 5 August 2013.

26. Donnelly ML, Luke G, Mehrotra A, Li X, Hughes LE, et al. (2001) Analysis of the aphthovirus 2A/2B polyprotein "cleavage" mechanism indicates not a proteolytic reaction, but a novel translational effect: a putative ribosomal "skip". J Gen Virol 82: 1013–1025 Available: http://www.ncbi.nlm.nih.gov/pubmed/11297676. Accessed 3 March 2014.

27. De Felipe P (2004) Skipping the co-expression problem: the new 2A "CHYSEL" technology. Genet Vaccines Ther 2: 13 Available: http://www.gvt-journal.com/content/2/1/13. Accessed 16 August 2013.

28. Kalderon D, Roberts BL, Richardson WD, Smith AE (1984) A short amino acid sequence able to specify nuclear location. Cell 39: 499–509 Available: http://dx.doi.org/10.1016/0092-8674(84)90457-4. Accessed 27 August 2013.

29. Morris GJ, Coulson GE, Leeson EA (1985) Changes in the shape of mitochondria following osmotic stress to the unicellular green alga Chlamydomonas reinhardii. J Cell Sci 76: 145–153 Available: http://www.ncbi.nlm.nih.gov/pubmed/3905835. Accessed 27 August 2013.

30. Ehara T, Osafune T, Hase E (1995) Behavior of mitochondria in synchronized cells of Chlamydomonas reinhardtii (Chlorophyta). J Cell Sci 108 (Pt 2: 499–507 Available: http://www.ncbi.nlm.nih.gov/pubmed/7768996. Accessed 27 August 2013.

31. Hiramatsu T, Nakamura S, Misumi O, Kuroiwa T (2006) Morphological Changes In Mitochondrial And Chloroplast Nucleoids And Mitochondria During The Chlamydomonas Reinhardtii (Chlorophyceae) Cell Cycle. J Phycol 42: 1048–1058 Available: http://doi.wiley.com/10.1111/j.1529-8817.2006.00259.x. Accessed 27 August 2013.

32. Harris EH, Stern DB, Witman GB (2009) The Chlamydomonas Sourcebook. Second Edi. Elsevier Available: http://dx.doi.org/10.1016/B978-0-12-370873-1.00002-2. Accessed 28 August 2013.

33. Fischer N, Rochaix JD (2001) The flanking regions of PsaD drive efficient gene expression in the nucleus of the green alga Chlamydomonas reinhardtii. Mol Genet Genomics 265: 888–894 Available: http://www.ncbi.nlm.nih.gov/pubmed/11523806. Accessed 27 August 2013.

34. Voeltz GK, Rolls MM, Rapoport TA (2002) Structural organization of the endoplasmic reticulum. EMBO Rep 3: 944–950 Available: http://www.pubmedcentral.nih.gov/articlerender.fcgi?artid=1307613&tool=pmcentrez&rendertype=abstract. Accessed 27 August 2013.

35. Gomord V, Denmat L-A, Fitchette-Laine A-C, Satiat-Jeunemaitre B, Hawes C, et al. (1997) The C-terminal HDEL sequence is sufficient for retention of secretory proteins in the endoplasmic reticulum (ER) but promotes vacuolar targeting of proteins that escape the ER. Plant J 11: 313–325 Available: http://doi.wiley.com/10.1046/j.1365-313X.1997.11020313.x. Accessed 27 August 2013.

36. De Hostos EL, Togasaki RK, Grossman A (1988) Purification and biosynthesis of a derepressible periplasmic arylsulfatase from Chlamydomonas reinhardtii. J Cell

Biol 106: 29–37 Available: http://www.pubmedcentral.nih.gov/articlerender.fcgi?art id=2114941&tool=pmcentrez&rendertype=abstract. Accessed 28 August 2013.

37. Schroda M (2004) The Chlamydomonas genome reveals its secrets: chaperone genes and the potential roles of their gene products in the chloroplast. Photosynth Res 82: 221–240 Available: http://www.ncbi.nlm.nih.gov/pubmed/16143837. Accessed 28 August 2013.

38. Kim JH, Lee S-R, Li L-H, Park H-J, Park J-H, et al. (2011) High cleavage efficiency of a 2A peptide derived from porcine teschovirus-1 in human cell lines, zebrafish and mice. PLoS One 6: e18556 Available: http://dx.plos.org/10.1371/journal.pone. 0018556. Accessed 12 August 2013.

39. Fang J, Qian J-J, Yi S, Harding TC, Tu GH, et al. (2005) Stable antibody expression at therapeutic levels using the 2A peptide. Nat Biotechnol 23: 584–590 Available: http://www.ncbi.nlm.nih.gov/pubmed/15834403. Accessed 3 February 2014.

40. Szymczak AL, Workman CJ, Wang Y, Vignali KM, Dilioglou S, et al. (2004) Correction of multi-gene deficiency in vivo using a single "self-cleaving" 2A peptide-based retroviral vector. Nat Biotechnol 22: 589–594 Available: http://www.ncbi.nlm.nih. gov/pubmed/15064769. Accessed 20 January 2014.

41. Keasling JD (2012) Synthetic biology and the development of tools for metabolic engineering. Metab Eng 14: 189–195 Available: http://www.sciencedirect.com/science/article/pii/S1096717612000055. Accessed 20 January 2014.

42. Halpin C (2005) Gene stacking in transgenic plants—the challenge for 21st century plant biotechnology. Plant Biotechnol J 3: 141–155 Available: http://www.ncbi.nlm. nih.gov/pubmed/17173615. Accessed 23 January 2014.

43. Ha S-H, Liang YS, Jung H, Ahn M-J, Suh S-C, et al. (2010) Application of two bicistronic systems involving 2A and IRES sequences to the biosynthesis of carotenoids in rice endosperm. Plant Biotechnol J 8: 928–938 Available: http://www.ncbi.nlm. nih.gov/pubmed/20649940. Accessed 4 February 2014.

44. Zhang F, Rodriguez S, Keasling JD (2011) Metabolic engineering of microbial pathways for advanced biofuels production. Curr Opin Biotechnol 22: 775–783 Available: http://www.sciencedirect.com/science/article/pii/S0958166911000875. Accessed 20 January 2014.

45. Heinig U, Gutensohn M, Dudareva N, Aharoni A (2013) The challenges of cellular compartmentalization in plant metabolic engineering. Curr Opin Biotechnol 24: 239–246 Available: http://www.ncbi.nlm.nih.gov/pubmed/23246154. Accessed 15 August 2013.

46. Sweetlove LJ, Fernie AR (2013) The spatial organization of metabolism within the plant cell. Annu Rev Plant Biol 64: 723–746 Available: http://www.ncbi.nlm.nih. gov/pubmed/23330793. Accessed 13 August 2013.

47. Dumas P, Bergdoll M, Cagnon C, Masson JM (1994) Crystal structure and sitedirected mutagenesis of a bleomycin resistance protein and their significance for drug sequestering. EMBO J 13: 2483–2492 Available: http://www.pubmedcentral. nih.gov/articlerender.fcgi?artid=395119&tool=pmcentrez&rendertype=abstract. Accessed 4 February 2014.

There are several supplemental files that are not available in this version of the article. To view this additional information, please use the citation on the first page of this chapter.

CHAPTER 10

Development of a Broad-Host Synthetic Biology Toolbox for *Ralstonia eutropha* and Its Application to Engineering Hydrocarbon Biofuel Production

CHANGHAO BI, PETER SU, JANA MÜLLER, YI-CHUN YEH, SWAPNIL R. CHHABRA, HARRY R. BELLER, STEVEN W. SINGER AND NATHAN J. HILLSON

10.1 INTRODUCTION

Chemoautotrophic "Knallgas" bacteria can utilize H_2/CO_2 for growth under aerobic conditions, and have great potential to directly produce liquid fuels from CO_2 and/or syngas [1,2]. *Ralstonia eutropha* (*R. eutropha*), the model bacterium of this class, can grow to very high cell densities (>200 g/L) [3]. Under nutrient limitation, *R. eutropha* directs most of its carbon flux to the synthesis of polyhydroxybutyrate (PHB), a biopolymeric compound stored in granules. Under autotrophic growth conditions with H_2/CO_2, *R. eutropha* has been reported to synthesize 61 g/L of PHB (representing ~70% of total cell weight) in 40 h [4]. With random mutagenesis and relatively simple engineering, PHB and related polyhydroxyalkanoate polymers have been produced in *R. eutropha* on industrial scales [3].

While *R. eutropha* has great potential to be engineered to produce desired compounds (beyond PHB) directly from CO_2, little work has been done to develop genetic part libraries to enable such endeavors. Although suicide vectors have been used to generate in-frame deletions and point mutations in *R. eutropha* [5], and previously reported broad-host range expression systems [6] may be transferable to *R. eutropha*, to date, the only established inducible expression system for *R. eutropha* has been a pBBR1-derived vector with a P_{BAD} promoter [7]. Here, we have initiated the development of a synthetic biology toolbox to enable complex metabolic engineering applications in *R. eutropha* H16. We evaluated a variety of vectors, promoters, 5' mRNA stem-loop sequences, and ribosomal binding sites (RBSs), and rationally mutated and engineered these genetic components to improve and diversify their function in *R. eutropha*. We then applied the resulting toolbox to engineer and optimize a hydrocarbon production pathway. Taken together, this work develops and demonstrates the engineering utility of a plasmid-based toolbox for *R. eutropha*.

10.2 RESULTS

10.2.1 BROAD-HOST VECTOR EVALUATION AND ENGINEERING

Three broad-host-range plasmid vectors were selected as starting points for the construction of new plasmid-based expression systems for *R. eutropha*: 1) pCM62, a low-copy-number plasmid within the IncP incompatibility group [8]; 2) pBBR1MCS, a medium-copy-number plasmid [9]; and 3) pKT230, a high-copy-number plasmid within the IncQ group containing the RSF1010 origin [10]. The kanamycin-resistance selection marker within pKT230 was replaced with a chloramphenicol-resistance marker to enable co-selection with pCM62 and pBBR1MCS-derivative plasmids. An inducible rfp expression cassette containing a P_{BAD} promoter [7], an E. coli consensus RBS, rfp, and a double terminator was incorporated into all three plasmid types. While none of the resulting plasmids were successfully electroporated into *R. eutropha*, they were all successfully transconjugated. As shown in Figure 1A, pBADrfp (the pBBR1MCS-derivative, see Table 1) provided the highest induced RFP expression level, while pK-Trfp and pCMrfp had lower expression levels. Plasmid pBADrfp (BBR1 origin, kanamycin resistance) co-propagated stably with pKTrfp (KT origin, chloramphenicol resistance) or pCMrfp (CM62 origin, tetracycline resistance) in *R. eutropha* (data not shown).

To increase the copy number of the pCMrfp plasmid, previously reported site-directed mutations were made to the trfA gene [15]. While putatively high-copy-number pCMrfp mutants (TrfA positions 251, 254 and

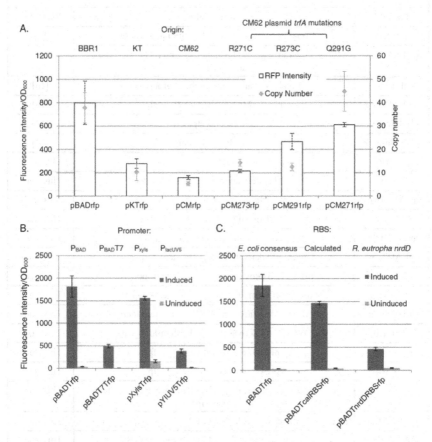

FIGURE 1: Plasmid expression vector RFP fluorescence intensities and copy numbers. (A) Induced RFP fluorescence intensity/OD$_{600}$ (dark bars) and plasmid copy numbers (square dots) for the various origins of replication. (B, C) Induced (dark bars) and uninduced (light bars) RFP fluorescence intensity/OD$_{600}$ for the various (B) promoters and (C) RBS sequences.

TABLE 1: Strains and plasmids used in this study.

Strain	Description	Reference
R. eutropha H16	*R. eutropha* wildtype strain currently classified as *Cuparividus necator*	ATCC 17669
R. eutropha H16 Δ2303	H16 Δ(H16_A0459-0464, H16_A1526-1531); Δbeta ox; mutant is deficient in native-oxidation	[11]
E. coli DH10B	F' proA+B+lacI^q ΔlacZM15/ fhuA2 Δ(lac-proAB) glnV galR (zgb-210::Tn10) Tet^{s}endA1 thi-1 Δ(hsdS-mcrB)5	NEB
E. coli S17	*E. coli* host strain for transconjugation	[12]

Plasmid	Description	Reference
pCM62	Broad host-range plasmid IncP group; amp^R, tet^R	[8]
pBBR1MCS	Broad host-range plasmid compatible with IncQ, IncP, IncW, and colE1; kan^R	[9]
pKT230	Broad host-range plasmid IncQ group; kan^R	[10]
pBADrfp	pBBR1MCS derivative; P$_{BAD}$_rfp	[13]
pBbE8c-RFP	colE1; P$_{BAD}$_rfp; cm^R	[14]
pBbA8a-RFP	p15a; P$_{BAD}$_rfp; amp^R	[14]
pKTrfp	pKT230 derivative; P$_{BAD}$_rfp; cm^R	This study
pCMrfp	pCM62 derivative; P$_{BAD}$_rfp; amp^R, tet^R	This study
pCM271rfp	pCMrfp with TrfA R271C mutation	This study
pCM273rfp	pCMrfp with TrfA R273C mutation	This study
pCM291rfp	pCMrfp with TrfA Q291G mutation	This study
pBADTrfp	pBADrfp derivative; P$_{BAD}$_T7 stem-loop_rfp	This study
pXylsTrfp	pBADTrfp derivative; P$_{xyls/PM}$_T7 stem-loop_rfp	This study
pUV5Trfp	pBADTrfp derivative; P$_{lacUV5}$_T7 stem-loop_rfp	This study
pIUV5Trfp	pBADTrfp derivative; lacI^q; P$_{lacUV5}$_T7 stem-loop_rfp	This study
pTetTrfp	pBADTrfp derivative; P$_{tet}$_T7 stem-loop_rfp	This study
pProErfp	pBADTrfp derivative; P$_{proE}$_rfp	This study
pProSrfp	pBADTrfp derivative; P$_{proS}$_rfp	This study
pBADT7Trfp	pBADTrfp derivative; P$_{BAD}$_T7 polymerase; PT7_T7 stem-loop_rfp	This study
pYIUV5Trfp	pBADTrfp derivative; lacY lacI^q; P$_{lacUV5}$_T7 stem-loop_rfp	This study
pBADTcalRB-Srfp	pBADTrfp derivative; P$_{BAD}$_T7 stem-loop_calRBS$_{rfp}$_rfp	This study
pBADTnrd-DRBSrfp	pBADTrfp derivative; P$_{BAD}$_T7 stem-loop_nrdDRBS_rfp	This study

TABLE 1: CONTINUED.

pKTTrfp	pKT230 derivative; $P_{BAD\_}$ T7 stem-loop_rfp; cm^R	This study
pCMTrfp	pCM62 derivative; $P_{BAD\_}$ T7 stem-loop_rfp	This study
pCM271Trfp	pCM62 derivative; pCMTrfp with TrfA R271C mutation	This study
pCM271Tcal-RBSrfp	pCM271rfp derivative; $P_{BAD\_}$T7 stem-loop_calRBS$_{rfp\_}$rfp	This study
pBADTHC	pBADrfpT derivative; $P_{BAD\_}$T7 stem-loop_aar_adc	This study
pKTTHC	pKTrfp derivative; $P_{BAD\_}$T7 stem-loop_aar_adc	This study
pCMTHC	pCM62 derivative; $P_{BAD\_}$T7 stem-loop_aar_adc	This study
pCM271THC	pCM271rfp derivative; $P_{BAD\_}$T7 stem-loop_aar_adc	This study
pBADHC	pBADrfp derivative; $P_{BAD\_}$aar_adc	This study
pXylsTHC	pBADrfp derivative; $P_{xyls/PM\_}$T7 stem-loop_aar_adc	This study
pYIUV5THC	pYIUV5Trfp derivative; $P_{lacUV5\_}$T7 stem-loop_aar_adc	This study
pBADTcalRB-SHC	pBADTHC derivative; $P_{BAD\_}$T7 stem-loop_calRBS$_{aar\_}$aar_cal-RBS$_{adc\_}$adc	This study
pBADTnrd-DRBSHC	pBADTHC derivative; $P_{BAD\_}$T7 stem-loop_nrdDRBS_aar_ nrd-DRBS_adc	This study
pCM271Tcal-RBSHC	pCM271rfp derivative; $P_{BAD\_}$T7 stem-loop_ calRBS$_{aar\_}$aar_cal-RBS$_{adc\_}$adc	This study

234) were not successfully transconjugated and established in *R. eutropha*, possibly because high-copy-number plasmids are not well tolerated in *R. eutropha*[16], medium-copy-number pCMrfp mutants (TrfA R271C, R273C and Q291G) were established. The mutant pCMrfp plasmids pCM271rfp, pCM273rfp, and pCM291rfp were measured to have higher RFP expression levels than pCMrfp (Figure 1A). To determine the absolute copy numbers of the pCMrfp plasmid variants, qPCR was performed using *R. eutropha* colonies as the source of the template (Figure 1A). pCM271rfp was determined to have the highest copy number (44.8 ± 8.5 copies per cell) among the pCMrfp variants. pCM273rfp and pCM291rfp both had higher copy numbers than pCMrfp.

10.2.2 T7 STEM-LOOP STRUCTURE EVALUATION

A T7 stem-loop structure [17] was inserted upstream of the RBS of the rfp gene on plasmid pBADrfp, yielding pBADTrfp. Introducing the T7 stem-loop structure into pBADTrfp increased RFP expression (1814±236 RFP intensity/OD, Figure 1B) by approximately 2-fold over pBADrfp levels (798±185 RFP intensity/OD) (Figure 1A).

10.2.3 INDUCIBLE PROMOTER SYSTEM EVALUATION AND ENGINEERING

In addition to PBAD, several other inducible promoter systems were evaluated in *R. eutropha*. Various repressor or activator genes along with their respective operators and promoters were inserted into pBADTrfp, replacing araC/PBAD. As shown in Figure 1B, the PBAD (pBADTrfp) and Pxyls/PM (pXylsTrfp) promoter systems provided the highest RFP expression upon induction. This is the first demonstration that the Pxyls/PM promoter system is functional in *R. eutropha*. The T7 promoter controlled by PBAD-induced T7 polymerase (pBADT7Trfp), although only providing modest RFP expression upon induction, had very little expression in the absence of induction. PlacUV5 (pUV5Trfp and pIUV5Trfp), Ptet (pTetTrfp), and Ppro (pProErfp and pProSrfp) systems did not show inducible expression in *R. eutropha* (Additional file 1: Figure S1, and data not shown). The Plac/lacI system has been reported previously not to be functional in *R. eutropha*[18]. Genomic sequence comparison between *R. eutropha* H16 and E. coli revealed that *R. eutropha* lacks the galactose permease gene lacY. This permease facilitates the transportation of lactose as well as the Plac inducer IPTG into E. coli[19]. A lacY gene codon-optimized for *R. eutropha* expressed from a constitutive promoter was incorporated into pIUV5Trfp, yielding pYIUV5Trfp. As shown in Figure 1B and Additional file 1: Figure S1, the incorporation of the lacY gene into pYIUV5Trfp enabled the IPTG-inducible expression of RFP from PlacUV5, although the expression level is low compared to those of PBAD and Pxyls/PM.

Cross-induction perturbation assays were performed to test if the chemical inducers L-arabinose (PBAD), m-toluic acid (Pxyls/PM), and IPTG (PlacUV5) affect the performance of their non-cognate promoter systems (Table 2). For the most part, the three chemical inducers did not significantly perturb their non-cognate promoter systems. For example, the induction of Pxyls/PM by 1 mM m-toluic acid retained 95.4% and 98.0% of normal levels, respectively, when 1 mM IPTG or 0.1% L-arabinose were added. An important exception is that 1 mM m-toluic acid negatively impacted the induction of PBAD by 0.1% L-arabinose to about 60% of normal levels. However, when the m-toluic acid concentration was reduced from 1 mM to 0.5 mM, the induction of PBAD by 0.1% L-arabinose remained at $91.7 \pm 1.6\%$ of normal levels.

10.2.4 RIBOSOMAL BINDING SITE SEQUENCE EVALUATION

Three RBS sequences were evaluated to compare their translation initiation efficiencies in *R. eutropha*: 1) an E. coli consensus RBS sequence (pBADTrfp), 2) a RBS calculator [20] designed RBS sequence (pBADT-calRBSrfp), and 3) the *R. eutropha* nrdD RBS sequence (pBADTnrd-DRBSrfp). RBS calculator parameters were specified towards designing a strong RBS sequence for *R. eutropha*, with the setting at "max", provided pBADTrfp RBS region context. The E. coli consensus RBS provided the highest RFP expression levels (Figure 1C), while the computationally designed RBS provided medium to high RFP expression, and the native *R. eutropha* nrdD RBS provided the lowest RFP expression levels.

TABLE 2: Promoter cross-induction test.

Promoter	Non-cognate inducer added		
	L-arabinose	m-toluic acid	IPTG
PBAD	(100%)[a]	$61.5 \pm 3.9\%$	$105.0 \pm 28.8\%$
Pxyls/PM	$98.0 \pm 2.7\%$	(100%)[a]	$95.4 \pm 9\%$
PlacUV5	$97.4 \pm 3.3\%$	$91.0 \pm 0.24\%$	(100%)[a]

[a]Cognate inducer alone.
Observed florescence intensity relative to cognate inducer alone.

10.2.5 APPLYING THE TOOLBOX TO HYDROCARBON PRODUCTION OPTIMIZATION

The synthetic biology toolbox was iteratively applied to optimize hydrocarbon production in *R. eutropha*. Genes encoding acyl-ACP reductase (aar) and aldehyde decarbonylase (adc) [21] were codon optimized for *R. eutropha* and synthesized (GenScript). These two synthesized genes were incorporated as an operon into the *R. eutropha* expression vectors developed above. The first set of constructed vectors (pBADTHC, pKT-THC, pCMTHC, and pCM271THC) was designed to determine the impact of plasmid origin of replication on hydrocarbon product titer (Figure 2A). Independent of the origin of replication, expressing the aar-adc hydrocarbon pathway in *R. eutropha* H16 resulted predominantly in the production of pentadecane (from palmitic acid) and heptadecene (likely from oleic acid [1]). The pBADTHC plasmid (pBBR1 origin) achieved the highest combined (pentadecane+heptadecene) titer, whereas the pCMTHC plasmid (pCM62 origin) produced the lowest. The pCM271THC plasmid (mutant pCM271 origin) was able to achieve a combined hydrocarbon titer comparable to that of pBADTHC, albeit with a more balanced pentadecane:heptadecene ratio. Removing the T7 stem-loop structure from pBADTHC, yielding plasmid pBADHC, did not significantly affect hydrocarbon titer, with the combined hydrocarbon for both reaching approximately 1000 µg/L (Additional file 1: Figure S2). The next set of constructed vectors (pXylsTHC and pYIUV5THC) was designed to determine the impact of the promoter on hydrocarbon product titer (Figure 2B). Of the three promoters tested, PBAD achieved the highest levels of hydrocarbon production, while Pxyls/PM and PlacUV5 only achieved low hydrocarbon titers (Figure 2B). The final set of constructed vectors (pBADTcalRBSHC and pBADTnrdDRBSHC) was designed to determine the impact of the RBS sequence on hydrocarbon product titer (Figure 2C). The E. coli consensus RBS sequence (pBADTHC) (tandem placement 5' of both aar and adc) achieved the highest combined hydrocarbon titer, while the calculated (calRBSaar_aar and calRBSadc_adc) and the *R. eutropha* nrdD (tandem placement 5' of both aar and adc) RBSs produced about 70% and 30% as much, respectively. The calculated RBSs achieved the most balanced pentadecane:heptadecene ratio. Since changing the

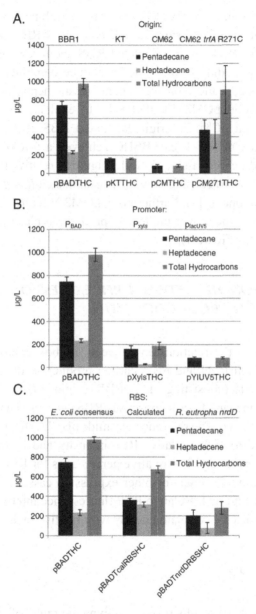

FIGURE 2: Hydrocarbon titers for the various (A) origins of replication, (B) promoters, and (C) RBS sequences; (dark bars) pentadecane, (light bars) heptadecene, and (grey bars) combined.

pBBR1 origin/E. coli consensus RBS sequence combination (pBADTHC) to either mutant pCM271 origin/E. coli consensus RBS sequence (pC-M271THC) or pBBR1 origin/calculated RBS sequence (pBADTcalRB-SHC) combinations did not dramatically reduce combined hydrocarbon titers, but produced a more balanced pentadecane:heptadecene ratio, we constructed plasmid pCM271TcalRBSHC to evaluate the hydrocarbon titer of the mutant pCM271 origin/calculated RBS sequence combination. Surprisingly, pCM271TcalRBSHC achieved a 6-fold improvement in combined hydrocarbon titer (~6 mg/L, Figure 3) relative to pBADTHC, the highest titer construct using previously established *R. eutropha* expression system components [7]. Furthermore, pCM271TcalRBSHC achieved a 100-fold improvement over the lowest production plasmids, pCMTHC and pYIUV5THC (Figure 4).

10.2.6 RELATIONSHIP BETWEEN RFP EXPRESSION LEVEL AND HYDROCARBON PRODUCTION TITER

To visually assess the relationship between toolbox component effects on RFP expression level and hydrocarbon production titer, the pBADT expression cassette (consisting of a pBBR1origin, PBAD promoter, T7 stemloop sequence, and E. coli consensus RBS) was used as a normalization point of reference. The reference plasmids pBADTrfp and pBADTHC were normalized to 100% relative RFP intensity and hydrocarbon production titer, respectively. The relative percentages for RFP intensity and hydrocarbon production titer for other expression cassette plasmid pairs are presented in Figure 4. Relative RFP fluorescence intensity appears to only slightly positively correlate linearly with relative hydrocarbon titer.

10.3 DISCUSSION

Our reporter construct results revealed a dynamic range of rfp expression levels for the genetic parts in the toolbox (Figure 1) and rapidly identified those parts that are non-functional in *R. eutropha* H16 Δ2303 (Additional file 1: Figure S1). We then investigated the impact of the toolbox on

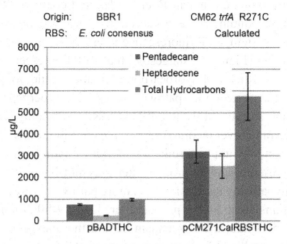

FIGURE 3: Hydrocarbon titers for the two highest producing strains.

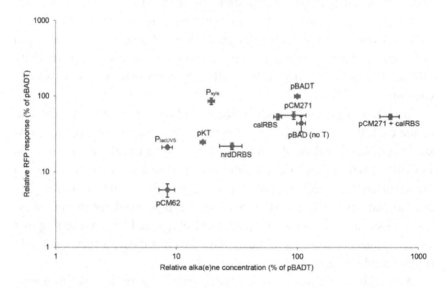

FIGURE 4: RFP expression levels and hydrocarbon production titers. Values normalized to reference plasmids pBADTrfp and pBADTHC. pCM62: pCMTrfp and pCMTHC. P_{lacUV5}: pYIUV5Trfp and pYIUV5THC. pKT: pKTTrfp and pKTTHC. P_{xyls}: pXylsTrfp and pXylsTHC. nrdDRBS: pBADTnrdDRBSrfp and pBADTnrdDRBSHC. calRBS: pBADTcalRBSrfp and pBADTcalRBSHC. pCM271: pCM271Trfp and pCM271THC. pBADT: pBADTrfp and pBADTHC. pBAD (no T): pBADrfp and pBADHC. pCM271+calRBS: pCM271TcalRBSrfp and pCM271TcalRBSHC.

hydrocarbon production titer in *R. eutropha*, and compared the resulting titers with corresponding rfp levels across the various expression configurations. In our system, RFP fluorescence intensity weakly correlated with hydrocarbon titer (Figure 4), suggesting that RFP fluorescence intensity is only a modest predictor of hydrocarbon titer, and should not generally be assumed to be a surrogate for pathway genes. This is perhaps not surprising, given that: 1) even RFP and GFP expression may only weakly correlate with each other over identical expression configurations [22], and thus RFP expression may not be a reliable reporter of hydrocarbon production pathway expression; and 2) product titer may not linearly or monotonically relate to pathway expression (i.e., more pathway expression does not necessarily translate to higher product titers [23]). While future work will be required to explain the mechanism behind the unexpected 6-fold improvement in combined hydrocarbon titer for pCM271TcalRBSHC over pBADTHC (Figures 3 and 4), we suspect that subtle differences in pathway expression may have serendipitously resulted in substantial titer increases. As such, what is more important than simply maximizing expression of pathway components is the capability to fine tune expression with sufficient granularity to resolve pathway bottlenecks and alleviate toxicity effects. The work reported here specifically contributes to this important capability.

We were surprised to observe that including the pCM271 vector and/or the calculated RBS parts in the hydrocarbon pathway expression construct resulted in balanced levels of pentadecane:heptadecane production in contrast with all other configurations for which the ratio was skewed predominantly to pentadecane (Figures 2 and 3). While the underlying mechanism for the relationship between pentadecane:heptadecane skew on expression configuration remains to be elucidated, it is interesting that using various components of the toolbox developed here affected not only overall product titers, but also product ratios.

Although we did not leverage this capability here, it is worth noting that since plasmid pBADrfp (BBR1 origin, kanamycin resistance) co-propagated stably with pKTrfp (KT origin, chloramphenicol resistance) or pCMrfp (CM62 origin, tetracycline resistance) in *R. eutropha*, it would be possible to engineer a multi-gene metabolic pathway across two plasmid vectors in the same cell (for example, see [24]). Furthermore, promoter cross-induction test results (Table 2) suggest that separate inducible promoters could be used to independently tune the expression of different portions of the pathway. These capabilities will play important future roles in engineering and optimizing more complex metabolic pathways in *R. eutropha*.

Metabolic engineering efforts often focus on a small set of microbial hosts, such as *E. coli* and *Saccharomyces cerevisiae*, simply because there are many established genetic and heterologous gene expression tools available for them. This select group of model microbes, however, may have limited utility for many industrial applications of interest. On the other hand, microbial hosts with metabolic capabilities and growth conditions well suited for specific industrial applications (like *R. eutropha*, which can function as a chemolithoautotroph), but with limited genetic tools, are extremely challenging and time-consuming to metabolically engineer, and developing new genetic tools for specific microbes of interest can be entire research efforts in and of themselves [25]. Here, we have developed and deployed the toolbox for the metabolic engineering of *R. eutropha*. From the outset, we designed our efforts with broad-host range applicability in mind so that we could readily apply the same tools to other microbial hosts of interest. For example, the RSF1010 origin-derived pKT plasmids developed here are able to replicate in a wide range of Gram-negative bacteria (e.g., Enterococci) as well as the phylogenetically distant cyanobacteria, which are also important hosts of interest for biofuels production [10]. Plasmids within the IncP incompatibility group (including pCM62) and the Pxyls/PM promoter system have been demonstrated to function in Pseudomonas putida[26,27]. All vectors reported here contain mob genes to bolster efficacy across a wide range of bacteria. We envision that the broad-host range toolbox developed here will serve as a turnkey foundation for developing a robust set of metabolic engineering tools for other microbes of interest by simplifying and streamlining the process of screening for functional expression systems that operate within the microbe of interest. Building upon this vision, the toolbox could be exploited to screen metabolic pathway performance across multiple microbial hosts, through the direct transfer of constructs (e.g., pCM271TcalRBSHC) to microbes with overlapping functional expression systems. This approach, especially when coupled with no or low leakage inducible promoters (e.g., PBADT7, Figure 1B), may be particularly effective for identifying microbial hosts that are tolerant to target or pathway intermediate compounds that are toxic to model microbes such as E. coli.

10.4 CONCLUSIONS

In this work, we have developed a toolbox for the metabolic engineering of *R. eutropha* H16, comprising six vectors spanning three compatibility groups, four promoter systems responding to three chemical inducers, a T7 5' mRNA stem-loop structure, and three RBSs. The major contribution of this work is that through increasing genetic regulatory part diversity, we have extended the dynamic range and tunable granularity of gene expression available for *R. eutropha*. We have demonstrated the value of the developed toolbox by increasing combined pentadecane and heptadecene hydrocarbon production titer 6-fold over that achievable with previously available gene expression tools and 100-fold over that achieved by our lowest producing engineered strains. Due to the broad-host range of the selected vectors and mobilized plasmid construction, this toolbox has a great potential to be applied to other microbial hosts for metabolic engineering purposes.

10.5 MATERIALS AND METHODS

10.5.1 BACTERIAL CULTIVATION

R. eutropha H16 (ATCC 17669), *R. eutropha* H16 Δ2303 [11], *E. coli* DH10B (NEB) and S17 [12] were grown at 30°C in lysogeny broth (LB). Kanamycin (50 mg/L for *E. coli*; 200 mg/L for *R. eutropha*), ampicillin (50 mg/L), chloramphenicol (30 mg/L), tetracycline (10 mg/L) and/or gentamycin (10 mg/L) were added to the medium as appropriate.

10.5.2 PLASMID CONSTRUCTION

With the exceptions of pBADTrfp, pBADTcalRBSrfp, pBADT7Trfp, pBADTnrdDRBSrfp, pXylsTrfp, pKTrfp, pCMrfp, pCM271rfp, pCM273rfp, and pCM291rfp (see Additional file 1), plasmids were assembled with the CPEC or Gibson methods [28,29], and corresponding DNA assembly protocols and DNA oligo primers were designed with j5 and DeviceEditor [30,31]. DNA templates were PCR-amplified with Phusion high-fidelity polymerase (Thermo Scientific). PCR products were gel purified before CPEC or Gibson assembly. The assembled plasmids were either transformed into E. coli DH10B, screened by colony PCR [32], sequence validated (Quintara Biosciences), and then transformed into E. coli S17 or directly transformed into *R. eutropha* H16 Δ2303, screened by colony PCR, and then sequence validated. Plasmids were then transconjugated from E. coli S17 into *R. eutropha* H16 Δ2303 as previously described [12].

10.5.3 STRAIN AND PLASMID AVAILABILITY

The strains and plasmids used in this study are listed in Table 1. All strains and plasmids developed here, along with their associated information (e.g., annotated GenBank-format sequence files, sequence validation trace files, DeviceEditor design files, and j5 design output files), have been deposited in the public instance of the JBEI Registry [33] (https://public-registry. jbei.org webcite; entries JPUB_001171-JPUB_001230) and are physically available from the authors and/or addgene (http://www.addgene.org webcite) upon request.

10.5.4 RFP FLUORESCENCE ASSAY

To measure the fluorescence intensity of RFP (monomeric mRFP1, maturation < 1 hour) [34] expressed from each type of plasmid vector, single colonies were picked and inoculated into LB seed-culture tubes supplemented with kanamycin, chloramphenicol, gentamycin, or tetracycline, as appropriate. 100 µL of each overnight seed culture was inoculated into a fresh 5 mL LB culture tube supplemented with the appropriate antibiotic and inducer (IPTG, L-arabinose, m-toluic acid, or tetracycline) and grown at 30°C, 200 rpm for 48 hours. 100 µL of each cell culture tube was then added to a separate well in a 96-well clear-bottom plate (Corning: No. 3631) and RFP fluorescence was measured with a Safire (Tecan) microplate reader using an excitation wavelength of 585 nm and an emission wavelength of 620 nm. OD600 was also measured for each well immediately thereafter to calculate the RFP fluorescence intensity/OD600 ratio reported for this assay.

10.5.5 PLASMID COPY NUMBER ASSAY

Single colonies (serving as templates) were picked, suspended in water, and then boiled for 10 minutes. Primers (pcmF and pcmR, phaZF and phaZR; Additional file 1) to amplify 400 to 500 bp fragments of rfp and phaZ respectively, were designed with Clone Manager 8.0. qPCR reactions were performed with a StepOnePlus Real-Time PCR System (Life Technologies) with Maxima SYBR Green/Fluorescein qPCR Master Mix (Thermo Scientific) as recommended by the manufacturers. Three biological replicates, with two technical replicates each, were performed for each plasmid type. Absolute plasmid copy numbers were determined using the CT difference between the plasmid-borne rfp gene and the single copy chromosomal phaZ gene.

10.5.6 INDUCER DOSE RESPONSE ASSAY

For each plasmid type, 10 µL of overnight LB cell culture was inoculated into each of 4 separate 24-well clear-bottom plate wells containing 1 mL fresh LB supplemented with a varying concentration of the appropriate inducer. These culture plates were then grown in a Pro200 (Tecan) microplate reader at 30°C, 37 rpm, for 72 hours. OD600 and RFP fluorescence intensity, using an excitation wavelength of 585 nm and an emission wavelength of 620 nm, were measured after 48 hours.

10.5.7 HYDROCARBON PRODUCTION ASSAY

Single colonies were picked and inoculated into 10 mL fresh LB glass culture tubes and grown at 30°C for 15 hours. The appropriate inducer was then added to each culture tube (final concentration: 1 mM IPTG, 0.1% L-arabinose, or 1 mM m-toluic acid). 1 mL decane (Sigma, 99% purity) was then immediately added to each 10 mL culture tube to extract hydrocarbons and other metabolites. 72 hours after induction, 100 μL of each decane overlay was removed for direct gas chromatography–mass spectrometry (GC/MS) analysis. Electron ionization (EI) GC/MS analyses were performed with a model 7890A GC quadrupole mass spectrometer (Agilent) with a DB-5 fused silica capillary column (J & W Scientific, 30-m length, 0.25-mm inner diameter, 0.25-μm film thickness) coupled to a HP 5975C mass selective detector. 1 μL injections were performed by an Agilent model 7683B autosampler. The GC oven was typically programmed to ramp from 40°C (held for 3 minutes) to 300°C at 15°C/min and then held for 5 minutes. The injection port temperature was 250°C, and the transfer line temperature was 280°C. The carrier gas, ultra-high purity helium, flowed at a constant rate of 1 mL/min. Injections were splitless, with the split turned on after 0.5 minutes. For full-scan data acquisition, the MS typically scanned from 50 to 600 atomic mass units at a rate of 2.7 scans per second. One of the major products, pentadecane (15:0), was quantified with m/z 57 area, while the other major product, heptadecene (17:1), was quantified with m/z 83 area of authentic standards (Sigma, 99% purity).

REFERENCES

1. Müller J, MacEachran D, Burd H, Sathitsuksanoh N, Bi C, Yeh YC, Lee TS, Hillson NJ, Chhabra SR, Singer SW, Beller HR: Engineering of Ralstonia eutropha H16 for autotrophic and heterotrophic production of methyl ketones. Appl Environ Microbiol 2013, 79:4433-4439.
2. Li H, Opgenorth PH, Wernick DG, Rogers S, Wu TY, Higashide W, Malati P, Huo YX, Cho KM, Liao JC: Integrated electromicrobial conversion of CO2 to higher alcohols. Science 2012, 335:1596.

3. Reinecke F, Steinbuchel A: Ralstonia eutropha strain H16 as model organism for PHA metabolism and for biotechnological production of technically interesting biopolymers. J Mol Microbiol Biotechnol 2009, 16:91-108.
4. Ishizaki A, Tanaka K, Taga N: Microbial production of poly-D-3-hydroxybutyrate from CO2. Appl Microbiol Biotechnol 2001, 57:6-12.
5. Lenz O, Friedrich B: A novel multicomponent regulatory system mediates H2 sensing in Alcaligenes eutrophus. Proc Natl Acad Sci USA 1998, 95:12474-12479.
6. Fodor BD, Kovacs AT, Csaki R, Hunyadi-Gulyas E, Klement E, Maroti G, Meszaros LS, Medzihradszky KF, Rakhely G, Kovacs KL: Modular broad-host-range expression vectors for single-protein and protein complex purification. Appl Environ Microbiol 2004, 70:712-721.
7. Delamarre SC, Batt CA: Comparative study of promoters for the production of polyhydroxyalkanoates in recombinant strains of Wautersia eutropha. Appl Microbiol Biotechnol 2006, 71:668-679.
8. Marx CJ, Lidstrom ME: Development of improved versatile broad-host-range vectors for use in methylotrophs and other Gram-negative bacteria. Microbiology-Sgm 2001, 147:2065-2075.
9. Kovach ME, Phillips RW, Elzer PH, Roop RM 2nd, Peterson KM: pBBR1MCS: a broad-host-range cloning vector. Biotechniques 1994, 16:800-802.
10. Sode K, Tatara M, Ogawa S, Matsunaga T: Maintenance of broad host range vector Pkt230 in marine unicellular cyanobacteria. Fems Microbiology Letters 1992, 99:73-78.
11. Brigham CJ, Budde CF, Holder JW, Zeng Q, Mahan AE, Rha C, Sinskey AJ: Elucidation of beta-oxidation pathways in Ralstonia eutropha H16 by examination of global gene expression. J Bacteriol 2010, 192:5454-5464.
12. Simon R, Priefer U, Puhler A: A broad host range mobilization system for invivo genetic-engineering - transposon mutagenesis in gram-negative bacteria. Bio/Technology 1983, 1:784-791.
13. Yeh YC, Muller J, Bi C, Hillson NJ, Beller HR, Chhabra SR, Singer SW: Functionalizing bacterial cell surfaces with a phage protein. Chem Commun (Camb) 2013, 49:910-912.
14. Lee TS, Krupa RA, Zhang F, Hajimorad M, Holtz WJ, Prasad N, Lee SK, Keasling JD: BglBrick vectors and datasheets: a synthetic biology platform for gene expression. J Biol Eng 2011, 5:12.
15. Durland RH, Toukdarian A, Fang F, Helinski DR: Mutations in the trfA replication gene of the broad-host-range plasmid RK2 result in elevated plasmid copy numbers. J Bacteriol 1990, 172:3859-3867.
16. Haugan K, Karunakaran P, Tondervik A, Valla S: The host range of RK2 minimal replicon copy-up mutants is limited by species-specific differences in the maximum tolerable copy number. Plasmid 1995, 33:27-39.
17. Mertens N, Remaut E, Fiers W: Increased stability of phage T7g10 mRNA is mediated by either a 5'- or a 3'-terminal stem-loop structure. Biol Chem 1996, 377:811-817.
18. Fukui T, Ohsawa K, Mifune J, Orita I, Nakamura S: Evaluation of promoters for gene expression in polyhydroxyalkanoate-producing Cupriavidus necator H16. Appl Microbiol Biotechnol 2011, 89:1527-1536.

19. Hansen LH, Knudsen S, Sorensen SJ: The effect of the lacY gene on the induction of IPTG inducible promoters, studied in Escherichia coli and Pseudomonas fluorescens. Curr Microbiol 1998, 36:341-347.
20. Salis HM, Mirsky EA, Voigt CA: Automated design of synthetic ribosome binding sites to control protein expression. Nat Biotechnol 2009, 27:946-950.
21. Schirmer A, Rude MA, Li X, Popova E, del Cardayre SB: Microbial biosynthesis of alkanes. Science 2010, 329:559-562.
22. Mutalik VK, Guimaraes JC, Cambray G, Lam C, Christoffersen MJ, Mai QA, Tran AB, Paull M, Keasling JD, Arkin AP, Endy D: Precise and reliable gene expression via standard transcription and translation initiation elements. Nat Methods 2013, 10:354-360.
23. Ma SM, Garcia DE, Redding-Johanson AM, Friedland GD, Chan R, Batth TS, Haliburton JR, Chivian D, Keasling JD, Petzold CJ, et al.: Optimization of a heterologous mevalonate pathway through the use of variant HMG-CoA reductases. Metab Eng 2011, 13:588-597.
24. Redding-Johanson AM, Batth TS, Chan R, Krupa R, Szmidt HL, Adams PD, Keasling JD, Lee TS, Mukhopadhyay A, Petzold CJ: Targeted proteomics for metabolic pathway optimization: application to terpene production. Metab Eng 2011, 13:194-203.
25. Zhao DL, Yu ZC, Li PY, Wu ZY, Chen XL, Shi M, Yu Y, Chen B, Zhou BC, Zhang YZ: Characterization of a cryptic plasmid pSM429 and its application for heterologous expression in psychrophilic Pseudoalteromonas. Microb Cell Fact 2011, 10:30.
26. Dammeyer T, Steinwand M, Kruger SC, Dubel S, Hust M, Timmis KN: Efficient production of soluble recombinant single chain Fv fragments by a Pseudomonas putida strain KT2440 cell factory. Microb Cell Fact 2011, 10:11.
27. Dammeyer T, Timmis KN, Tinnefeld P: Broad host range vectors for expression of proteins with (Twin-) Strep-tag, His-tag and engineered, export optimized yellow fluorescent protein. Microb Cell Fact 2013, 12:49.
28. Quan J, Tian J: Circular polymerase extension cloning for high-throughput cloning of complex and combinatorial DNA libraries. Nat Protoc 2011, 6:242-251.
29. Gibson DG, Young L, Chuang RY, Venter JC, Hutchison CA 3rd, Smith HO: Enzymatic assembly of DNA molecules up to several hundred kilobases. Nat Methods 2009, 6:343-345.
30. Hillson NJ, Rosengarten RD, Keasling JD: j5 DNA assembly design automation software. ACS Synth Biol 2012, 1:14-21.
31. Chen J, Densmore D, Ham TS, Keasling JD, Hillson NJ: DeviceEditor visual biological CAD canvas. J Biol Eng 2012, 6:1.
32. Linshiz G, Stawski N, Poust S, Bi C, Keasling JD, Hillson NJ: PaR-PaR laboratory automation platform. ACS Synth Biol 2013, 2:216-222.
33. Ham TS, Dmytriv Z, Plahar H, Chen J, Hillson NJ, Keasling JD: Design, implementation and practice of JBEI-ICE: an open source biological part registry platform and tools. Nucleic Acids Res 2012, 40:e141.
34. Campbell RE, Tour O, Palmer AE, Steinbach PA, Baird GS, Zacharias DA, Tsien RY: A monomeric red fluorescent protein. Proc Natl Acad Sci USA 2002, 99:7877-7882.

PART V

NANOTECHNOLOGY
AND CHEMICAL ENGINEERING

CHAPTER 11

Heterogeneous Photocatalytic Nanomaterials: Prospects and Challenges in Selective Transformations of Biomass-Derived Compounds

JUAN CARLOS COLMENARES AND RAFAEL LUQUE

11.1 BACKGROUND

Developing artificial systems that can mimic natural photosynthesis to directly harvest and convert solar energy into usable or storable energy resources has been the dream of scientists for many years. The use of solar energy to drive organic syntheses is indeed not a novel concept. The idea was originally proposed by Ciamician as early as 1912. [1] However, the common use of the generally accepted term photocatalysis and significant developments in this field properly started in the 1970s after the discovery of water photolysis on a TiO_2 electrode by Fujishima and Honda. [2]

Heterogeneous photocatalytic nanomaterials: prospects and challenges in selective transformations of biomass-derived compounds. ©*The Royal Society of Chemistry 2014.* Chemical Society Reviews, *2014, 43, 765 (DOI: 10.1039/c3cs60262a). Creative Commons Attribution 3.0 Unported Licence (http://creativecommons.org/licenses/by/3.0/).*

The energy crisis provided a strong impulse to research alternative energy sources, hoping that mankind could mimic nature by exploiting solar energy for the generation of fuels (e.g. hydrogen via water splitting). [2,3] Additionally, the pollution concerns and the increasing demand for more sustainable sources of chemicals also prompted the search for alternative solutions potentially able to clean up water and air avoiding the addition of further chemicals and to explore new "green" routes for chemicals production. A third later application in chemical synthesis was strongly related to the fact that the same photocatalysts were applied. [4] Very important is the fact that all highlighted three applications can be grouped into green and sustainable processes.

Photocatalysis, in which solar photons are used to drive redox reactions to produce chemicals (e.g. fuels), is the central process to achieve this aim. Despite significant efforts to date, a practically viable photocatalyst with sufficient efficiency, stability and low cost is yet to be demonstrated. It is often difficult to simultaneously achieve these different performance metrics with a single material component. The ideal heterogeneous photocatalysts with multiple integrated functional components could combine individual advantages to overcome the drawbacks of single component photocatalysts. For further information, readers are kindly referred to recent overviews on the development of advanced photocatalysts. [4,5]

As a fundamental and applied field of science, heterogeneous photochemistry continues to be an important component of modern chemistry in the 21st century. Research in this field has significantly evolved during the last three decades (especially titanium oxide chemistry) with enhanced knowledge on mechanisms, development of new technologies for storage and conversion of solar energy, environmental detoxification of liquid and gaseous ecosystems, and the photochemical production of new materials. More recently, a new research avenue related to selective transformations of biomass and residues to high added value products has emerged as a potentially useful alternative to conventional heterogeneously catalyzed processes. This contribution is aimed to provide an overview of recent work conducted along the lines of selective photochemical transformations, particularly focused on photocatalysis for lignocellulose-based biomass valorization. Future prospects and work in progress in this field will also be emphasized. The effective utilization of clean, safe, and abundant solar energy is envisaged to provide energy, chemicals as well as solving environmental issues in the future and an appropriate semiconductor-photoinduced biomass/waste conversion can be the key for such transformations.

11.2 FUNDAMENTAL PRINCIPLES OF SEMICONDUCTOR PHOTOCATALYSIS

The fundamental principles of heterogeneous photocatalysis have been extensively reported in previous reports. [5] A photocatalytic transformation is initiated when a photoexcited electron is promoted from the filled valence band (VB) of a semiconductor photocatalyst (e.g., TiO_2) to the empty conduction band (CB) as the absorbed photon energy, hv, equals or exceeds the band gap of the semiconductor photocatalyst. As a consequence, an electron and hole pair ($e^- - h^+$) are formed (eqns (1)–(4)).

$$\text{Photon activation: } TiO2 + hv \rightarrow e^- + h^+ \quad (1)$$

$$\text{Oxygen adsorption: } (O_2)ads + e- \rightarrow O_2^{\cdot -} \quad (2)$$

$$\text{Water ionization: } H_2O \rightarrow OH^- + H^+ \quad (3)$$

$$\text{Superoxide protonation: } O_2^{\cdot -} + H^+ \rightarrow HOO^{\cdot} \quad (4)$$

$HOO^{\cdot}$ radicals (eqn (4)) also have scavenging properties similar to oxygen, thus prolonging the photohole lifetime (eqns (5) and (6)).

$$HOO^{\cdot} + e^- \rightarrow HO_2^- \quad (5)$$

$$HOO^- + H^+ \rightarrow H_2O_2 \quad (6)$$

Redox processes can take place at the surface of the photoexcited photocatalyst. Very fast recombination between electron and hole occurs unless oxygen (or any other electron acceptor) is available to scavenge the elec-

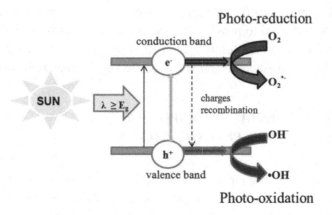

FIGURE 4: Photo-activation of a semiconductor and primary reactions occurring on its surface.

trons to form superoxides ($O_2^{\cdot-}$) (Fig. 1), hydroperoxyl radicals ($HO_2^{\cdot}$) and subsequently H_2O_2.

In contrast to conventional catalysis thermodynamics, not only spontaneous reactions ($\Delta G < 0$) but also non-spontaneous reactions ($\Delta G > 0$) can be promoted by photocatalysis. The input energy is used to overcome the activation barrier in spontaneous reactions so as to facilitate photocatalysis at an increased rate or under milder conditions. In comparison, for non-spontaneous processes, part of the input energy is converted into chemical energy that is accumulated in the reaction products.

The generated holes have high potential to directly oxidize organic species (the mechanism is not definitively proven) or indirectly via the combination with ˙OH abundant in water solution (eqns (7)–(9)). [5a,b,6]

$$H_2O + h^+ \rightarrow OH^{\cdot} + H^+ \qquad (7)$$

$$R\text{–}H + OH^{\cdot} \rightarrow R^{\cdot} + H_2O \qquad (8)$$

$$R^{\cdot} + h^+ \rightarrow R^{+\cdot} \rightarrow \text{Mineralization products} \qquad (9)$$

The corresponding mineral acid of the non-metal substituent is formed as a by-product (e.g. in organic waste maybe a different heteroatom including Cl, N, S) (eqn (10)).

$$\text{Organic waste} => \text{photocatalyst/O}_2/\text{ hv} \geq \text{Eg} => \text{Intermediate(s)}$$
$$=> CO_2 + H_2O + \text{Mineral acid} \qquad (10)$$

11.2.1 REACTIVE OXYGEN SPECIES (ROS) FORMED DURING A PHOTOCATALYTIC PROCESS

Several highly reactive ROS can oxidize a large variety of organic pollutants in heterogeneous photocatalysis. These include

˙OH (redox potential +2.81 V vs. standard hydrogen electrode, SHE). ˙OH radicals act as a main component during photo-degradation reactions, particularly for substances that have weak affinity to the TiO_2 surface. [5a,b] ˙OH can be produced from the oxidation of surface hydroxyls or adsorbed water. However, recent studies have pointed out that the role of ˙OH radicals is probably underestimated. For many organic compounds, the primary one-electron oxidation should be initiated by free or trapped holes.

O2˙⁻ (redox potential +0.89 V vs. SHE)/HO₂˙. Superoxide anions (O2˙⁻), easily protonated to yield HO_2˙ in acidic solution (pK_a = 4.8), are readily generated from molecular oxygen by capturing photoinduced electrons. O2˙⁻ species are generally less important in initiating oxidation reactions but mainly participate in the total mineralization of organic compounds via reaction with organoperoxy radicals and production of H_2O_2 (redox potential +1.78 V vs. SHE) by O_2˙⁻ disproportionation.

O₃⁻. O_3^- species are generated in reactions between the photoformed hole center on the lattice oxygen (OL⁻) and molecular oxygen.

¹O₂ singlet molecular oxygen. ¹O₂ is usually formed via energy transfer from the triplet state of a dye to molecular oxygen. It has been suggested that oxidation of O_2˙⁻ by holes (redox potential +2.53 V vs. SHE) at the TiO_2 surface should be a plausible mechanism to produce ¹O₂.

Apart from the important role of ROS in photocatalytic degradation processes, the control and generation of ROS is essential in heterogeneous photocatalysis to be potentially able to design and predict pathways in selective organic photocatalytic oxidations.

11.3 CHEMICALS FROM LIGNOCELLULOSIC BIOMASS

Lignocellulose is a highly complex and rather recalcitrant feedstock comprising three major polymeric units: lignin, cellulose and hemicellulose. Cellulose (a high molecular weight polymer of glucose, ca. 45 wt%) forms bundles that are additionally attached together by hemicellulose (ca. 20 wt%). Cellulose and hemicellulose are surrounded by lignin (less than 25 wt%) which provides extra rigidity and recalcitrance to the entire material. Lignocellulosic feedstocks (i.e., forestry waste, agricultural residues, municipal paper waste and certain food waste residues) can be converted into a variety of useful products in multi-step processes. [7,8] However, due to the large complexity of such feedstocks, lignocellulosics have to be broken down by several processes and technologies into simpler fractions which can be converted into desired products. These include major routes such as gasification, pyrolysis and pre-treatment hydrolysis/fragmentation steps. Upon deconstruction, the obtained solid, liquid or gaseous fractions need to be upgraded via various processes which yield a plethora of chemicals. [7,8]

From a sustainability viewpoint, the development of low-temperature, highly selective (photo)catalytic routes for the direct transformation of lignocellulosics into valuable chemicals or platform molecules (e.g. glucose, carboxylic acids) is of great significance. [7] These compounds can then be subsequently converted into useful products. However, the selective conversion of lignocellulosics under mild conditions still remains a significant challenge owing to the high recalcitrant structure of cellulose and lignin fractions. For further information on topics related to cellulose and lignin depolymerization, readers are kindly referred to recent literature overviews. [9–11] Subsequent sections have been aimed to illustrate the potential of photocatalytic processes for the conversion of selected lignocellulosic feedstocks into chemicals and fuels.

11.4 PHOTOCATALYTIC REFORMING OF LIGNOCELLULOSE-BASED ORGANIC WASTE: HYDROGEN PRODUCTION

Biomass sources have been utilized for sustainable hydrogen production. [12] A number of processes have been developed for this purpose which include fast pyrolysis, supercritical conversion and steam gasification and many others which however require harsh reaction conditions (e.g., high temperatures and/or pressures) and consequently imply high costs. [12] Photocatalytic reforming may be a promising alternative as mild reaction conditions driven by solar light (i.e., room temperature and atmospheric pressure) can be comparatively advantageous to such energy-consuming thermochemical processes.

Hydrogen production by photocatalytic reforming of lignocellulosic biomass may also be more feasible and practical as compared to photocatalytic water-splitting due to its potentially higher efficiency. [13] The thermodynamics of photochemical water splitting were reported to store a maximum of only 12% of the incident light energy. [13]

Lignocellulosic biomass-derived compounds can also serve as sacrificial agents (electron donors) to reduce the photocatalyst recombination e^-–h^+ rate. A large variety of organic compounds (most of them model compounds of lignocellulose structure, e.g. alcohols, polyols, sugars, as well as organic acids) have been used as electron donors for photocatalytic hydrogen production. For further information, readers are kindly referred to recent accounts of the most pioneering studies by leading authors on biomass (photo)catalytic reforming for chemical (hydrogen included) production. [14]

11.5 HETEROGENEOUS PHOTOCATALYSIS FOR SELECTIVE CHEMICAL TRANSFORMATIONS

The utilization of heterogeneous photocatalysis for environmental purification (both in the liquid and gas phase) has been extensively investigated as photoactivated semiconductors have proven activities to unselectively mineralize various types of toxic, refractory and non-biodegradable organic pollutants under mild conditions. [15] Closely related to selective synthesis, structured photocatalytic systems have also been employed for the oxofunctionalization of hydrocarbons via selective oxidations (Fig. 2). [16]

Reported organic photosynthetic reactions include oxidations, reductions, isomerizations, substitutions, polymerizations and condensations. These reactions can be carried out in inert solvents and/or their combinations with water. For more details on photocatalytic synthetic transformations based on different photo-active solid materials, readers are kindly referred to leading reviews in the field. [17]

The following fundamental requirements should be met by any photocatalytic system harvesting and converting solar energy into chemical energy:

(a) the photoresponse of the system should optimally match the solar spectrum;

(b) photoexcited charges must be efficiently separated to prevent recombination;

(c) charges should have sufficient energy to carry out the desired chemical reactions (e.g. selective oxidations);

(d) the photocatalyst must be photo-stable, chemically and biologically inert and of low cost.

For photon transfer optimization, decisive operational parameters include light intensity, nature and concentration of the substrate, nature and concentration of the photocatalyst, pH of the solution, reaction temperature, type of the reactor, presence of oxygen and the content of ions. [18]

11.5.1 PHOTOCATALYTIC SELECTIVE OXIDATIONS OF CELLULOSE-BASED CHEMICALS

Catalytic oxidations have traditionally been carried out in environmentally harmful chlorinated organic solvents at high temperatures and pressures by employing stoichiometric amounts of various inorganic oxidants as oxygen donors (e.g. chromate and permanganate species). Such oxidants are expensive and toxic and they also produce large amounts of hazardous waste, therefore needing to be replaced by safer systems.

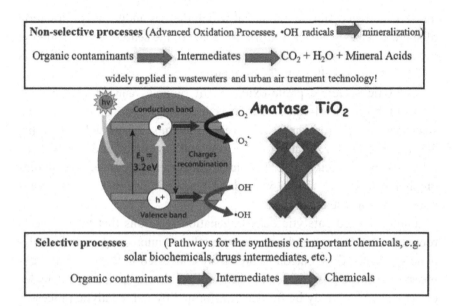

FIGURE 2: Production of different target compounds by semiconductor photo-induced catalytic routes.

Comparably, a photocatalytic process can bring about significant benefits in terms of milder and more environmentally sound reaction conditions and better selectivity for the desired product. Substances unstable at high temperatures may be synthesized via selective light-assisted processes.

Alcohol oxidation to their corresponding ketones, aldehydes, and/or carboxylic acids is one of the most important transformations in organic synthesis.19 Semiconductor photocatalysts have not been frequently employed in selective synthetic oxidation processes as the replacement of traditional oxidation methods has been mostly covered with heterogeneously catalyzed systems.

According to thermodynamics, an alcohol molecule with singlet electronic configuration cannot directly react with an unactivated dioxygen molecule, which has a triplet electronic configuration. [20] The working mechanism in metal oxidation catalysts involves an electron transfer-mediated step by the metal which thereby induces the formation of singlet oxygen species. [20] In this context, the search for an oxidation photocatalyst capable of directly activating dioxygen under solar light is an interesting as well as challenging task.

Catalytic selective photo-oxidation of biomass can provide a wide range of high added-value chemicals including some of the so-called sugar-derived platform molecules (e.g., succinic, 2,5-furandicarboxylic, 3-hydroxypropionic, gluconic, glucaric and levulinic acids as well as 3-hydroxybutyrolactone). [21] Table 1 summarizes recent selected studies on photocatalytic selective oxidation of lignocellulose-based model compounds to valuable chemicals.

Photo-assisted catalytic dehydrogenation reactions that take place at room temperature and ambient pressure offer an interesting route for aldehyde synthesis. C_1–C_4 alcohols are easily converted in the liquid and gas phase and in the presence of oxygen into their corresponding aldehydes or ketones, which may be further transformed by non-catalytic processes into acids. [22,23]

A range of different titania-based systems synthesized through sol–gel processes by varying the precursor and/or the ageing conditions (magnetic stirring, ultrasound, microwave or reflux) were recently reported for the liquid-phase selective photo-oxidation of crotyl alcohol to crotonaldehyde. The gas-phase selective photooxidation of 2-propanol to acetone was also chosen as a model reaction (Table 1, entry 3). [23] Both reactions showed relatively similar results in terms of influence of precursor and metal, despite having very different reactant/catalyst ratios and contact times. Titanium isopropoxide provided better results as compared to those achieved with titanium tetrachloride. The presence of iron, palladium or zinc in the systems was found to be detrimental for the activity. Zirconium and particularly gold improved the results as compared to pure titania (Table 1, entry 3).

A higher reactivity has been generally reported for primary alcohols [22] (Table 1, entry 2), which opens up a promising selective synthesis method for the production of hydroxycarboxylic acids. The selective dehydrogenation of various secondary alcohols was also possible. If the alcohol is unsaturated, isomerization may occur, yielding the corresponding saturated aldehyde.

The photocatalytic oxidation of alcohols is highly dependent on the type of alcohol. Generally, the conversion per pass of primary alcohols is low (with a slightly higher value for secondary alcohols), but high selectivities are generally achieved (>95%). The initial step of the proposed mechanism is the interaction of a surface hole with the hydroxyl group of the alcohol, forming metal-oxo species with proton removal. This proton removal step becomes easier with increasing carbon chain length and branching. The higher the number of adjacent hydrogen atoms present, the simpler the removal and thus improved conversions can be achieved.

Aliphatic carboxylic acids can be transformed to shorter chain acids (e.g. malic to formic acid) [24] or decarboxylated to the corresponding reduced hydrocarbons or hydrocarbon dimers in the absence of oxygen and in pure aqueous or mixed aqueous/organic solutions by means of photo-Kolbe-type processes (Table 1, entry 5). [25] As an example, the selective aqueous conversion of malic acid into formic acid has been conducted under visible light irradiation using a magnetically separable TiO_2-guanidine-(Ni,Co)-Fe_2O_4 nanocomposite (Table 1, entry 4). The photocatalyst featured a simple magnetic separation and offered the possibility to work under visible light and sunlight irradiation due to titania modification with guanidine, which remarkably decreased the band gap of the metal oxide semiconductor. [24] Comparably, acetic, propionic, butanoic and n-pentanoic acids could be decarboxylated to hydrocarbons in the absence of oxygen. For the case of acetic acid (a common product of biological processes), the aforementioned photo-Kolbe reaction could be combined with biological waste treatments to generate combustible fuels.

The efficiency of heterogeneous nano-TiO_2 catalysts in the selective photocatalytic oxidation of glucose into high-valued organic compounds has also been recently reported (Fig. 3). [26] This reaction was found to be highly selective (>70%) towards two organic carboxylic acids, namely glucaric (GUA) and gluconic acids (GA). These carboxylic acids are important building blocks for pharmaceutical, food, perfume or fuel industries. [18] CO_2 and traces of light hydrocarbons were also detected in the gas phase. Among all photocatalytic systems tested, the best product selectivity was achieved with titania synthesized by an ultrasound-modified sol–gel methodology (TiO_2(US)). [26] Solvent composition and short illumination times were proved to have a considerable effect on photocatalysts

TABLE 1. Selected samples of selective photocatalytic production of high value chemicals from lignocellulosic- biomass-derived organic molecules.

Entry	Model biomass-based molecule	Photocatalyst synthesis	Reaction conditions	Photocatalytic behavior (activity/selectivity)	Ref.
1	Methanol	Anatase-type TiO_2 particles (ST–01) having a large surface area of 300 $m^2 g^{-1}$	UV light (UV intensity: 1800 $mW\ cm^2$). Reaction in gas phase in the presence of air with a fixed bed of the catalyst	Highly selective (91 mol%) photooxidation of methanol to methyl formate with no catalyst deactivation. The conversion of methanol increased with the increase of the reaction temperature up to 250 1C (conversion is three times higher than that at room temperature but the selectivity decreases)	29
2	Ethanol, 1-propanol, 2-propanol, 2-butanol	TiO_2 rutile phase (3.6 and 7.7 $m^2 g^{-1}$)	Catalyst suspensions irradiated with 366 nm UV light	Selective production of ethanol, propanol and propanone with 0.101, 0.104 and 0.101 quantum yields at 20 1C, respectively	22
3	2-Propanol crotyl alcohol	TiO_2, Me–TiO_2 (Me = Pd, Pt, Zr, Fe, Zn, Ag, Au). Catalysts prepared by ultrasound- and microwave-assisted sol–gelprocedures	UV light (λ_{max} = 365 nm). Reaction in liquid phase (for crotyl alcohol) with a solid catalyst in suspension. Reaction in gas phase (for 2-propanol) with a fixed bed of the catalyst. Water used for cooling was thermostated at 10°C for reactions with crotyl alcohol and 20°C for 2-propanol	For crotyl alcohol, conversions between 8% and 38% for t = 30 min or 32% and 95% for t = 300 min for the two extreme catalysts (TiO_2:Fe and TiO_2:Au, respectively). When the influence of the metal is considered, iron, palladium and zinc exhibit lower conversions than the corresponding bare-titania, whereas the presence of silver, zirconium and especially gold is beneficial to photoactivity In the case of 2-propanol, platinumcontaining solids showed quite high selectivity values to acetone (in the 78–80% range at 22–28% conversion)	23
4	Malic acid	TiO_2-guanidine-(Ni,Co)-Fe_2O_4	Catalysts suspended in malic acid aqueous solution and illuminated under visible light (150 W Quartz Halogen Lamp, λ > 400 nm)	Selectivity close to 80% for formic acid could be achieved in less than 2 hours of reaction. Efficient separation of the photocatalyst after reaction	24

TABLE 1. CONTINUED.

5	Acetic, propionic, n-butanoic and npentanoic acids	Pt/TiO$_2$ (rutile). Pt was deposited by illuminating each powdered semiconductor suspended in water–ethanol solution	30 mL of water–organic acid (6 : 1 v/v) mixture and catalyst in suspension irradiated with a 500 W Xe lamp, pH < 2.0	In the absence of oxygen, aliphatic carboxylic acids (especially C$_4$–C$_5$ acids) are decarboxylated to the corresponding reduced hydrocarbons: acetic (156 micromol RH/10 h), propionic (1470 micromol RH/10 h), n-butanoic (996 micromol RH/10 h), n-pentanoic (1018 micromol RH/10h) acids	25
6	Glucose	Bare-TiO$_2$ and supported titania on zeolite Y. Catalysts were synthesized by a modified ultrasound-assisted sol–gel method	Catalysts suspended in a glucose solution (50%H2O/50%CH$_3$CN composition) and illuminated with a 125 W mercury lamp (λ_{max} = 365 nm)	High photoselectivity for glucaric and gluconic acid production (68.1% total selectivity, after 10 min. illumination time) especially for TiO$_2$ supported on zeolite Y. Apart from photocatalyst properties, it was found that reaction conditions, especially solvent composition and short illumination times, also have considerable effect on the activity/ selectivity of tested photocatalysts	26 and 27
7	Glucose	Fe-TiO$_2$ and Cr-TiO$_2$ supported on zeolite Y (SiO$_2$:Al$_2$O$_3$ = 80) and prepared by an ultrasoundassisted wet impregnation method (rotary evaporator was coupled with ultrasonic bath)	Catalysts suspended in a glucose solution (50%H$_2$O/50%CH$_3$CN solvent composition) and illuminated with a 125 W mercury lamp (λ_{max} = 365 nm)	Fe-TiO$_2$ zeolite-supported systems total selectivity for GUA+GA is 94.2% after 20 min of illumination Cr-TiO$_2$ zeolite-supported systems total selectivity for GUA+GA is 99.7% after 10 min of illumination	28
8	Lignocellulose	Commercial TiO$_2$	Solid state mixture of titania and lignocellulose UV irradiated (λmax = 360 nm)	Photocatalysis pretreatment used for an efficient biological saccharification of lignocellulosic materials. Titania did not disturb the biological reactions by cellulase and yeast	36

TABLE 1. CONTINUED.

| 9 | Kraft lignin | Ta_2O_5-IrO_2 thin film (prepared using a thermal decomposition technique) as electrocatalyst and TiO_2 nanotube (prepared using electrochemical anodization) as photocatalyst | Photochemical-electrochemical process in liquid phase. The intensity of the UV light was ca. 20 mW cm^{-2} (main line of emission, 365 nm) | Oxidation of lignin gave vanillin and vanillic acid | 37 |

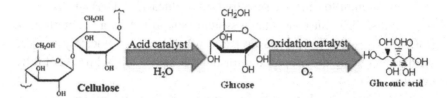

FIGURE 3: Acid catalytic hydrolysis of cellulose followed by catalytic partial oxidation of glucose to gluconic acid.

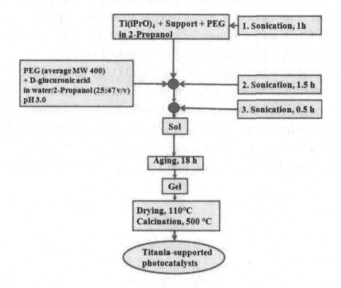

FIGURE 4: Schematic chart of TiO_2 supported on different materials (e.g., zeolite, silica) prepared by an ultrasound-assisted sol–gel methodology. [27]

activity/selectivity. Total organic compound selectivity was found to be 39% and 71% for liquid phase reactions using 10%Water/90%Acetonitrile and 50%Water/50%Acetonitrile, respectively. These values were obtained using the optimum TiO_2(US) photocatalyst.

Such results suggested that synthesized nano-TiO_2 material could in principle be used in the decomposition of waste from the food industry with the simultaneous production of high-value chemicals when residues (here glucose) act as electron donors. [18]

In an attempt to improve selectivities to glucaric (GUA) and gluconic (GA) acids, different nano-titania systems supported on a zeolite type Y (SiO_2:Al_2O_3 = 80) have been prepared. [27] These photocatalysts were synthesized using the proposed ultrasound-assisted sol–gel protocol fully detailed in Fig. 4.

Homogeneously distributed TiO_2 on zeolite Y provided an improved selectivity for glucose oxidation towards glucaric acid (GUA) and gluconic acid (GA). Total selectivity was ca. 68% after 10 min illumination time using a 1[thin space (1/6-em)]:[thin space (1/6-em)]1 H_2O–acetonitrile solvent composition. Results were comparably superior to those of unsupported TiO_2 and commercially available Evonik P-25 photocatalyst (Table 1, entry 6). Importantly, the more acidic character of the zeolite in comparison to silica-supported, bulk TiO_2(US) and Evonik P-25 may relate to the higher selectivity to carboxylic acids. Moreover, Y type zeolites are negatively charged materials which adsorb cationic substrates and repulse anionic analogues by electrostatic attraction. This behavior can facilitate the selective photocatalytic oxidation of glucose, as carboxylic acids formed under the reaction conditions (pH lowered from 6–7,

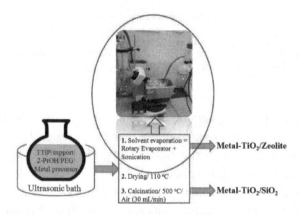

FIGURE 5: General procedure to synthesize metal-containing TiO_2 supported on different materials (e.g., zeolite, silica) by a sonication-assisted impregnation method (titanium tetraisopropoxide, TTIP; polyethylene glycol 400 molecular weight, PEG). [28]

depending on glucose solvent composition, to about 4–5 after 4 h of photocatalytic reaction) are repulsed, preventing at the same time their complete photo-oxidation/mineralization.

Further photocatalyst optimization via development of transition-metal containing supported nanotitania materials (Fig. 5) [28] provided advanced systems incorporating Fe or Cr able to achieve improved selectivities to carboxylic acids. No metal leaching (Fe, Cr, Ti) was detected after photoreaction, with Fe–TiO_2 systems being most selective (94% after 20 min of illumination under similar conditions previously reported, Table 1, entry 7).

The gas-phase selective photo-oxidation of methanol to methyl formate is another interesting process. [29] The reaction was carried out in a flow-type reactor to avoid deep oxidation of methanol in the presence of TiO_2 particles and under UV light irradiation. A high selectivity to methyl formate was observed (91%) with no catalyst deactivation observed under the investigated conditions (Table 1, entry 1).

The oxidation mechanism, efficiency and selectivity of the photocatalytic oxidation of alcohols using TiO_2 Evonik P25 were reported to strongly depend on the nature of the dispersing medium. Shiraishi et al. [16] and Morishita et al. [30] demonstrated the positive influence of acetonitrile on epoxide formation in the selective photo-oxidation of alkenes. The addition of small amounts of water to CH_3CN strongly inhibited alcohol adsorption and its subsequent oxidation as evidenced by ESR-spin trapping investigations. [31] The reactivity of alcohols on the surface of photoexcited TiO_2 was also found to be affected by the nature of their hydrophobic aliphatic chain. [31] Molecules including geraniol and citronellol were observed to be more susceptible to water content as compared to shorter chain analogues such as trans-2-penten-1-ol and 1-pentanol. Optimum reaction conditions were achieved in the photocatalytic oxidation of geraniol, citronellol, trans-2-penten-1-ol, and 1-pentanol to the corresponding aldehydes with good selectivity (>70%).

Along these lines, a partial photooxidation of diols including 1,3-butanediol, 1,4-pentanediol and vicinal-diols (e.g. 1,2 propanediol) could also be effectively conducted using TiO_2 in dichloromethane. [32] CO_2 was not observed as a reaction product but the observed main reaction products included two hydroxy-carbonyl compounds for each 1,n-diol. 1,2-propanediol mainly gave rise to hydroxyacetone (90% selectivity),

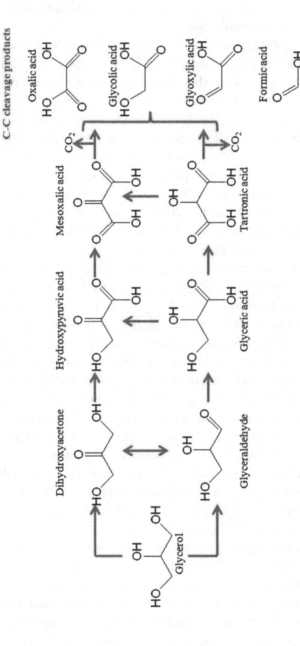

FIGURE 6: Pathways for glycerol photo-oxidation to valuable chemicals.

with only traces of pyruvic acid. Comparatively, 1,3-butanediol was converted into 3-hydroxybutyraldehyde and 4-hydroxy-2-butanone while 1,4-pentanediol gave 4-hydroxypentanal (75% selectivity) and certain quantities of 3-acetyl 1-propanol.

Similarly, glycerol is a relevant polyol currently produced in large quantities as a by-product of the biodiesel industry. The mechanism of the selective photocatalytic oxidation of glycerol was also recently reported in the presence of TiO_2 Evonik P25 and Merck TiO_2 (Fig. 6). [33] The product distribution observed at low glycerol concentration (glyceraldehyde and dihydroxyacetone) changes after a sharp maximum giving formaldehyde and glycolaldehyde as main products for P25 Evonik (mechanism derived from a direct electron transfer). Interestingly, mainly glyceraldehyde and dihydroxyacetone were observed on Merck TiO_2 ($^{\cdot}$OH-based mechanism), a material characterized by a lower density and more uniform population of hydroxyl groups at surface sites. These findings suggested that photocatalyst structures can significantly influence product distribution in photo-assisted processes. This example is particularly relevant to engineering of valuable products from a biofuel-derived by-product.

A comprehensive investigation on advanced mechanistic aspects of polyol photooxidation on photoactivated metal oxides was subsequently reported aiming to better understand the interaction of sugars with the surface of some semiconductor materials. [34] The authors demonstrated that carbohydrates are oxidized at sites involved in the formation of oxo bridges between the chemisorbed carbohydrate molecule and metal ions at the oxide surface (e.g. TiO_2, α-FeOOH, and α-Fe2O3). Bridging was claimed to inhibit the loss of water, promoting a rearrangement that leads to elimination of formyl radicals. For natural carbohydrates, the latter reaction mainly involves carbon-1, whereas the main radical products of the oxidation were observed to be radicals arising from H atom loss centered on carbon-1, -2, and -3 sites. Photoexcited TiO_2 oxidizes all carbohydrates and polyols, whereas α-FeOOH oxidizes some of the carbohydrates, α-Fe_2O_3 being unreactive. [34]

Detailed mechanistic studies on photocatalytic oxidation of related organic compounds (e.g., methane to methanol, partial oxidation of geraniol to produce only citral, glycerol oxidation and propene epoxidation among others) are obviously out of the scope of the present contribution, but readers are kindly referred to a recently reported comprehensive overview on this subject. [35]

11.5.2 LIGNOCELLULOSICS PHOTO-ASSISTED PRETREATMENTS AND SELECTIVE PHOTO-OXIDATIONS OF AROMATIC ALCOHOLS AS LIGNIN MODEL COMPOUNDS

Lignocellulosics pretreatment using (bio)chemical methodologies has attracted a great deal of attention in recent years. Enzymatic saccharification (SA) and subsequent fermentation (FE) are conventionally utilized processes for bio-ethanol production from cellulose materials. Importantly, an efficient biological saccharification of lignocellulosic material requires certain pre-treatment steps. Several reported pre-treatments include the use of dilute H_2SO_4, alkali or pressured hot water and even the utilisation of ionic liquids. Photocatalytic pre-treatments of lignocellulosics have however rarely been reported in the literature. Interestingly, Yasuda et al. [36] have recently devised a 2 step process for biological bioethanol production from lignocellulose via feedstock pre-treatment with titania photocatalyst under UV-illumination. Selected lignocellulosic feedstocks were napiergrass (Pennisetum purpureum Schumach) and silver grass (Miscanthus sinensis Andersson) as summarized in Table 1, entry 8. The photocatalytic pre-treatment did not significantly affect the final product distribution, demonstrating that TiO_2 was not interfering with biological reactions promoted by cellulases and yeast. Most importantly, the photocatalytic pre-treatment was remarkably effective in reducing the time in SA and FE reactions with respect to untreated and NaOH pre-treated feedstocks. The photocatalytic pretreatment was also proved to be an environmentally sound replacement for acid- and alkali-based processes. This report constitutes the first finding on photocatalytic pre-treatments featuring important insights into bioethanol production from cellulose.

Another example of the combination of a photocatalytic process with biomass oxidation comprised an integration of photochemical and electrochemical oxidation processes for the modification and degradation of kraft lignin applied (Table 1, entry 9).37 $Ta_2O5–IrO_2$ thin films were used as electrocatalysts with TiO_2 nanotube arrays as photocatalyst. Lignin deconstruction provided vanillin and vanillic acid, relevant compounds with important applications in the food and perfume industries.

In spite of these recent reports that exemplify the potential of photocatalysis for lignocellulosics conversion, selective photocatalytic transformations of lignocellulosic feedstocks are rather challenging due to the complex structure of lignocellulose, particularly related to the highly recalcitrant lignin fraction. Comparatively, relevant research has been conducted on model compounds which have been selected on the basis of representing structural motifs present in lignocellulose.

Aromatic alcohols have been used as model lignin compounds to be converted into various compounds via photocatalytic processing (Table 2). Among selective oxidation processes of alcohols, the conversion of benzyl alcohol to benzaldehyde is particularly attractive. Benzaldehyde is in fact the second most important aromatic molecule (after vanillin) used in the cosmetics, flavor and perfumery industries. Synthetic benzaldehyde is industrially produced via benzylchloride hydrolysis derived from toluene chlorination or through toluene oxidation. In this regard, alternative protocols able to selectively produce benzaldehyde from benzyl alcohol are in demand. While most research on selective production of benzaldehyde has been reported via heterogeneously catalyzed protocols, there are some relevant examples of such reactions under photo-assisted conditions. Aerobic aqueous suspensions of Au supported on cerium(IV) oxide (Au/CeO_2) were reported to be able to quantitatively oxidize benzyl alcohol to benzaldehydes under green light irradiation for a reaction time of 36 h (Table 2, entries 4–7). [38] Comparatively, the oxidation of benzyl alcohol was performed using differently prepared TiO_2 catalysts, reaching selectivities three- to seven fold superior to those of commercial TiO_2 catalysts. [39]

In terms of benzyl alcohol derivatives, the aqueous phase photocatalytic oxidation of 4-methoxybenzyl alcohol was achieved by using uncalcined brookite TiO_2. A maximum selectivity of ca. 56% towards aldehydes (i.e. ca. 3 times higher than that obtained with commercial TiO_2) was reported.

Heteropolyoxometalate (POM) catalysts of the type $[S_2M_{18}O_{62}]4+$ (M = W, Mo)[40] have also been developed to carry out the photo-oxidation of aromatic alcohols under sunlight and UV/Vis light in acetonitrile. Mechanistic investigations of photocatalysis by these nontoxic compounds indicate that near-UV/Vis irradiation of a POM solution results in an oxygen-to-metal charge-transfer excited state that has strong oxidation ability, responsible for the oxidation of organic substrates. Photoexcited POMs were observed to be reduced by the transfer of one or two electron(s) from the organic substrate. Other types of POMs including water tolerant materials by combining homogeneous POMs with photoactive and inactive supporting materials (e.g., mesoporous molecular sieves, titania in amorphous phase) and silica-encapsulated $H_3PW_{12}O_{40}$ have also been reported in similar chemistries.

The acceleration of the aerobic photo-oxidation of similar alcohols has also been described to take place on TiO_2 or SiO_2/TiO_2 samples without any loss of selectivity via surface loading of Bronsted acids (reaction time 2–7 h, conversion of the starting molecules ca. 42–100 mol%, selectivity 73–100 mol%). [41] The effect of Bronsted acids was confirmed and enhanced when a small quantity of SiO_2 was incorporated into TiO2 after the acid pretreatment, due to the presence of more acid sites on the TiO2 surface. The acceleration effect was attributed to the protons of Bronsted acids that effectively promote the decomposition of the formed surface peroxide-Ti species. This was demonstrated by means of in situ spectroscopy titration measurements which indicated that this phenomenon leads to surface site regeneration, contributing to an enhancement of the cycle efficiency without any loss of selectivity.

A similar selective photocatalytic oxidation of aromatic alcohols to aldehydes was systematically studied under visible-light irradiation utilizing anatase TiO_2 nanoparticles. [42] The unique features of the protocol rely on the possibility to use visible light irradiation apart from UV due to the formation of surface complex species by the adsorbed aromatic alcohol. Studies using fluorinated TiO_2 pointed to a dramatic decrease of photocatalytic activity under visible light irradiation which suggested a significant role played by the adsorption of the aromatic alcohol species on the surface of the solid involving the surface OH species. [42] Fig. 7 summarizes the reaction pathway proposed by Higashimoto et al.42a for the selective photocatalytic oxidation of benzyl alcohol in the presence of oxygen and titania. The surface complexes formed by the close interaction of the alcohol group or the aromatic ring with the hydroxyl surface group are able to absorb in the visible region, and hence the photoexcited complex undergoes hydrogen abstraction in the CH_2OH group by the

TABLE 2: Selected representative examples of selective photocatalytic oxidation of aromatic alcohols as model lignin compounds.

Entry	Reactants	Products	Conversion [%]	Selectivity [%]
Immobilized TiO_2 on silica cloth. Reaction conditions: alcohol, 1.43 mmol min^{-1}; O_2/alcohol = 22; 2.5-L annular photoreactor (average contact time 32 s); temperature = 190°C; conversion after 2-h illumination (UV light) time. [49]				
1	CH₂OH	CHO	35	>95
2	OH CH–CH₃	O C–CH₃	97	7 (83% styrene)
3	CH₂–CH₂OH	CH₂–CHO	53	26
Oxidation over 1 wt% Au/CeO2 under green light. Reaction conditions: benzyl alcohols 33 mmol; Au/CeO$_2$ 50 mg; 5 mL water; 1 bar O$_2$; 25°C; LED 530 nm; conversion after 20-h illumination time. [38]				
4	CH₂OH	CHO	>99	>99
5	CH₂OH CH₃	CHO CH₃	>99	>99
6	CH₂OH CH₃	CHO CH₃	>99	>99
7	CH₂OH H₃C	CHO H₃C	>99	>99
Carbon Quantum Dots (CQDs) under NIR light irradiation. Reaction conditions: 10 mmol alcohol; H$_2$O$_2$ (10 mmol, 30 wt% in water); 8 mg of CQD catalyst; 60°C; NIR irradiation for 12 h. [46]				
8	CH₂OH	CHO	92	100

TABLE 2: CONTINUED.

	Substrate	Product		
9	CH_2OH benzyl alcohol, H_3C–	CHO, H_3C–	88	>99
10	CH_2OH, H_3CO–	CHO, H_3CO–	86	>99
11	OH, CH–CH_3 (phenyl)	O, C–CH_3 (phenyl)	90	>96
12	C–O–CH_2, OH (phenyl)	C–O–CHO (phenyl)	88	>99

$Au_2(DP_{673})$/P25 catalyst. Reaction conditions: 5 mL toluene (solvent); 10 mmol alcohol; 20 mg catalyst; 1 bar O_2; 30°C; 4 h irradiation time (98% of light involves the visible range). [47]

	Substrate	Product		
13	OH, CH–CH_3 (phenyl)	O, C–CH_3 (phenyl)	79	100
14	CH_2OH (phenyl)	CHO (phenyl)	85	93
15	CH_2OH, H_3C–	CHO, H_3C–	>99	>100
16	CH_2OH, H_3CO–	CHO, H_3CO–	>99	91
17	OH, CH–CH_3, H_3C–	O, C–CH_3, H_3C–	83	99

TABLE 2: CONTINUED.

18			>99	>100
19			81	99

Multi-Pd core–CeO$_2$ shell nanocomposite. Reaction conditions: 20 h of visible light (λ > 420 nm) irradiation time. [48]

20			28	100
21			10	100
22			12	100

photoproduced holes forming a water molecule. The organic radical also releases one electron to form benzaldehyde. At the same time the photoinduced electrons are trapped by oxygen in the reacting medium.

Benzyl alcohol and some of its derivatives (para-structures) could be converted into their corresponding aldehydes at high conversions and selectivities (ca. 99%) under both UV and visible light irradiation. The only exception to this behavior was 4-hydroxybenzyl alcohol which was oxidized to 4-hydroxy benzaldehyde (selectivity of ca. 23% at ca. 85% conversion) along with some unidentified products. The authors claimed that OH groups from the TiO$_2$ surface reacted with the aromatic compound to form Ti–O–Ph species, exhibiting strong absorption in the visible region by ligand to metal charge transfer. These findings confirmed the results reported by Kim et al. [43] which proposed a direct electron transfer from the surface complexes to the conduction band of the TiO_2 upon absorbing visible light. Importantly, results evidenced that visible light can induce reactions by substrate–surface complexation enabling visible light absorption. Similar observations were recently reported by Li and co-workers [44] with nanorods of rutile titania phase synthesized by a hydrothermal reaction using rutile TiO_2 nanofibers obtained from calcination of composite electrospun nanofibers. The authors proposed a tentative mechanism where selective photocatalytic activities were proposed to be due to the visible-light absorption ability of the benzyl alcohol–TiO_2 nanorod complex and the unique properties of rutile TiO_2 nanorods. Excellent properties of these materials including a high surface-to-volume ratio, unidirectional 1D channels and superior survivability of electrons may contribute to more efficient electron transport, further required for benzaldehyde formation (>99% selectivity to benzaldehyde).

Visible light illumination of nitrogen-doped TiO_2 prepared by a sol–gel method brings about the selective oxidation of benzyl and cinnamyl alcohols to the corresponding aldehydes. Upon UV light irradiation, the alcohol conversion rate is faster but the reaction is unselective with respect

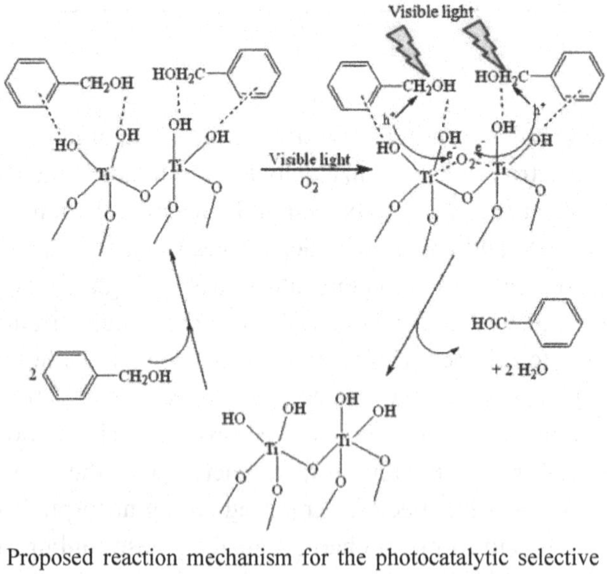

FIGURE 7: Proposed reaction mechanism for the photocatalytic selective oxidation of benzyl alcohol under visible light irradiation. [42][a]

to aldehyde formation. [45] The reaction takes place in oxygenated dry nitrile solvents (CH_3CN as the optimum solvent) and is totally inhibited in the presence of water. Alcohols were proved to be weakly adsorbed (do not readily capture the photogenerated holes) and the interaction of acetonitrile with the surface was demonstrated by its quenching of luminescence from band gap energy levels introduced by nitrogen. The formation of active oxygen species, particularly superoxide radicals, was proposed to be a key channel leading to aldehyde formation. Nevertheless, the possibility of dry acetonitrile playing an active role in the reaction mechanism cannot be ruled out as acetonitrile has been previously reported to promote the formation of active radical species. [45]

A new class of carbon nanomaterials known as Carbon Quantum Dots (CQDs) with sizes below 10 nm can also function as an effective near infrared (NIR) light driven photocatalyst for the selective oxidation of benzyl alcohol to benzaldehyde. [46] Based on the NIR light driven photo-induced electron transfer property and its photocatalytic activity for H_2O_2 decomposition, this metal-free catalyst could efficiently provide high yields of benzaldehyde under NIR light irradiation (Table 2, entries 8–12). Hydroxyl radicals were the main active oxygen species in benzyl alcohol selective oxidative reaction as confirmed by terephthalic acid photoluminescence probing assay, selecting toluene as a substrate. Such metal-free photocatalytic system also selectively converts other alcohol substrates to their corresponding aldehydes with high conversion, demonstrating a potential application of accessing traditional alcohol oxidation chemistry.

Plasmonic photocatalysts (noble metal – semiconductor hybrids) have been reported for selective formation of organic molecules. [47] Nano-sized (<5 nm) gold nanoparticles supported on anatase–rutile interphase (Evonik P-25 photocatalyst) by a deposition–precipitation technique could take advantage of plasmonic effects to achieve the photocatalytic selective production of several aromatic aldehydes from their corresponding aromatic alcohols (Table 2, entries 13–19). This photocatalysis can be promoted via plasmon activation of the Au particles by visible light followed by consecutive electron transfer in the Au/rutile/anatase interphase contact. Activated Au particles transfer their conduction electrons to rutile and then to adjacent anatase which catalyzes the oxidation of substrates by the positively charged Au particles along with reduction of oxygen by the conduction band electrons on the surface of anatase titania. Oxygen peroxide species have been found to be responsible for alcohol oxidation. $Au_2(DP_{673})$/P25 produced 17 µmol of acetophenone in the dark as a result of the high activity of the small Au particles. Light irradiation further enhanced the reaction to produce 74 µmol of acetophenone (over 4 times that obtained in the absence of light). Visible light illumination of Au/P25 catalyst clearly enhances the oxidation process.

Recent progress in metal core-semiconductor shell nanohybrids has also demonstrated that these systems can be potentially utilized as photocatalysts for selective organic oxidations. [48] A (Pt)–CeO_2 nanocomposite has been prepared in an aqueous phase with tunable core–shell and yolk–shell structure via a facile and green template-free hydrothermal approach, as illustrated in Fig. 8A. [48] The authors found that the core–shell nanocomposite could serve as an efficient visible-light-driven photocatalyst for the selective oxidation of benzyl alcohol to benzaldehyde using molecular oxygen as a green oxidant. The same authors also prepared various hybrids based on other metals (e.g. Pd) and carried out photocatalytic selective oxidation of benzylic alcohols over the multi-Pd core@CeO_2 shell nanocomposite (Table 2, entries 20–22). Electron–hole pairs are produced under visible light irradiation. Electrons are then trapped by the Pd cores and adsorbed benzyl alcohol interacts with holes to form the corresponding radical cation. Further reaction with dioxygen or superoxide radical species will lead to the formation of the corresponding aldehydes, with the plausible mechanism illustrated in Fig. 8B.

These photocatalytic protocols account for lab scale systems with relatively restricted applications. However, the possibility of scaling up photocatalytic processes to selective large scale processes has also been attempted in recent years. With regard to photocatalytic selective oxidation processes, various aromatic alcohols could be converted in a gas-phase 2.5-liter annular photoreactor using immobilized TiO_2 catalyst and under UV light (Table 2, entries 1–3). [49] The system was found to be specifically suited for the selective oxidation of primary and secondary aliphatic alcohols to their corresponding carbonyl compounds. Benzylic alcohols gave higher conversions, however, with more secondary reaction products. The presence of oxygen was found to be critical for the photooxidation. One disadvantage of the system relates to catalyst deactivation, which is attributed to the surface accumulation of reaction products. However, the catalyst could be regenerated by calcination in air for 3 h at 450 °C, providing similar activities to those of fresh catalysts. [49]

Another similar scale-up protocol was developed in a solar pilot plant reactor using TiO_2/Cu(II) photocatalyst. [50] A maximum yield of 53.3% for benzaldehyde could be obtained with respect to initial benzyl alcohol concentration (approx. 63% selectivity, reaction time 385 min), operating with an average temperature of 38 °C. It is well known that the substitution of oxygen with potential reducing species (in this case metal ions dissolved in the solution) still enables the oxidation of the organic species present in the solution. The authors claimed that in the absence of oxygen, one of the photochemical pathways is inhibited, leading to unselective OH radical production. In fact, the formation of superoxide radicals ($O2^{\cdot-}$) or hydrogen peroxide (H_2O_2) is not possible in the absence of oxygen, whose photolysis can generate OH radicals. In this case, OH radicals should form just as a result of the reaction of water with positive holes.

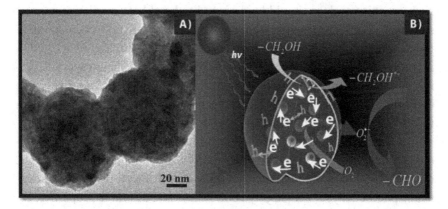

FIGURE 8: (A) TEM image of as-synthesized multi-Pd-core–CeO_2-shell nanoparticles; (B) a plausible mechanism for the selective oxidation of aromatic alcohols to aldehydes in the presence of multi-Pd-core–CeO_2-shell photocatalyst under visible light irradiation (adapted from ref. 48, Royal Society of Chemistry).

11.6 FUTURE CHALLENGES AND PROSPECTS

The concept to utilize solar light energy at room temperature and atmospheric pressure via photocatalysis to selectively convert lignocellulosic biomass to important chemicals and valuable products is a highly innovative approach that can bring several benefits from energy and environmental viewpoints. As such, photocatalytic selective processes are able to keep selectivities to products at reasonable levels at increasing conversions in the systems, as opposed to conventional thermally activated heterogeneous catalysis.

Photochemical reactions also require milder conditions to those generally present in thermal processes which may allow the conception of short and efficient reaction sequences, minimizing side processes making use of sunlight as a completely renewable source of energy (leaving no residue). In the near future, integrated systems of simultaneous photocatalyst-mediated chemical evolution and water purification may be possible. These processes could then be efficiently coupled to scale-up facilities (e.g. solar plants) able to provide continuous flow equipment and large reactors for important advances in the field.

Ideal photocatalysts for commercial applications in the valorization of such organic residues should be designed to be able to maximize conversion with a compromised selectivity for the desired product under solar light irradiation in aqueous solution (or a solventless environment), avoiding complete mineralization. However, even though most reported syntheses can be virtually carried out under sunlight, there are only very few reports on sunlight utilization for such transformations due to reproducibility issues in measurements.

This contribution has been aimed to illustrate that novel nanotechnology and nanomaterials design are currently bringing innovative concepts, ideas and protocols to the design and preparation of well-defined heterogeneous photocatalysts with highly controllable and desirable properties. By combining these new synthetic routes with surface-science, fundamentals of heterogeneous photocatalysis and theoretical mechanistic studies, we should possibly develop a new generation of highly stable and selective photocatalysts for oxidative organic transformations including those of lignocellulosic derivatives from which we hope to actively participate in future years to come.

REFERENCES

1. G. Ciamician, Science, 1912, 36, 385.
2. A. Fujishima and K. Honda, Nature, 1972, 238, 37.
3. J. A. Turner, Science, 2004, 305, 972.
4. S. Rawalekar and T. Mokari, Adv. Energy Mater., 2013, 3, 12.
5. (a) M. Fagnoni, D. Dondi, D. Ravelli and A. Albini, Chem. Rev., 2007, 107, 2725; (b) Y. Qu and X. Duan, Chem. Soc. Rev., 2013, 42, 2568 RSC; (c) H. Kisch, Angew. Chem., Int. Ed., 2013, 52, 812.
6. A. Fujishima, T. N. Rao and D. A. Tryk, J. Photochem. Photobiol., C, 2000, 1, 1.
7. M. Stoecker, Angew. Chem., Int. Ed., 2008, 47, 9200.
8. C. S. K. Lin, L. A. Pfaltzgraff, L. Herrero-Davila, E. B. Mubofu, S. Abderrahim, J. H. Clark, A. Koutinas, N. Kopsahelis, K. Stamatelatou, F. Dickson, S. Thankappan, Z. Mohamed, R. Brocklesby and R. Luque, Energy Environ. Sci., 2013, 6, 426.
9. A. Stark, Energy Environ. Sci., 2011, 4, 19.
10. (a) S. Van de Vyver, J. Geboers, P. A. Jacobs and B. F. Sels, ChemCatChem, 2011, 3, 82 CrossRef CAS Search PubMed; (b) C.-H. Zhou, X. Xia, C.-X. Lin, D.-S. Tong and J. Beltramini, Chem. Soc. Rev., 2011, 40, 5588 RSC; (c) J. Hilgert, N. Meine, R. Rinaldi and F. Schüth, Energy Environ. Sci., 2013, 6, 92 RSC.
11. (a) Q. Song, F. Wang, J. Cai, Y. Wang, J. Zhang, W. Yu and J. Xu, Energy Environ. Sci., 2013, 6, 994 RSC; (b) P. Gallezot, Chem. Soc. Rev., 2012, 41, 1538 RSC; (c) J. Zakzeski, P. C. A. Bruijincx, A. L. Jongerius and B. M. Weckhuysen, Chem. Rev., 2010, 110, 3552.
12. M. Ni, D. Y. C. Leung, M. K. H. Leung and K. Sumathy, Fuel Process. Technol., 2006, 87, 461.
13. J. R. Bolton, S. J. Strickler and J. S. Connolly, Nature, 1985, 316, 495.
14. Producing Fuels and Fine Chemicals from Biomass Using Nanomaterials, ed. R. Luque and A. M. Balu, Taylor and Francis Book Inc., New Jersey, 2013.
15. A. Di Paola, E. García-López, G. Marcì and L. Palmisano, J. Hazard. Mater., 2012, 211–212, 3.
16. Y. Shiraishi, M. Morishita and T. Hirai, Chem. Commun., 2005, 5977 RSC.
17. (a) M. D. Tzirakis, I. N. Lykakis and M. Orfanopoulos, Chem. Soc. Rev., 2009, 38, 2609 RSC; (b) C. Gambarotti, C. Punta, F. Recupero, T. Caronna and L. Palmisano, Curr. Org. Chem., 2010, 14, 1153.
18. J. C. Colmenares, R. Luque, J. M. Campelo, F. Colmenares, Z. Karpinski and A. A. Romero, Materials, 2009, 2, 2228.
19. S. E. Davis, M. S. Ide and R. J. Davis, Green Chem., 2013, 15, 17.
20. R. Ho, J. Liebman and J. Valentine, in Active Oxygen in Chemistry, ed. C. Foote, J. Valentine, A. Greenberg and J. Liebman, Springer, Berlin, 1995, pp. 1–23.
21. J. J. Bozell and G. R. Petersen, Green Chem., 2010, 12, 539 RSC.

22. P. R. Harwey, R. Rudham and S. Ward, J. Chem. Soc., Faraday Trans. 1, 1983, 79, 2975.
23. F. J. Lopez-Tenllado, A. Marinas, F. J. Urbano, J. C. Colmenares, M. C. Hidalgo, J. M. Marinas and J. M. Moreno, Appl. Catal., B, 2012, 128, 150.
24. A. M. Balu, B. Baruwati, E. Serrano, J. Cot, J. Garcia-Martinez, R. S. Varma and R. Luque, Green Chem., 2011, 13, 2750.
25. T. Sakata, T. Kawai and K. Hashimoto, J. Phys. Chem., 1984, 88, 2344.
26. J. C. Colmenares, A. Magdziarz and A. Bielejewska, Bioresour. Technol., 2011, 102, 11254.
27. J. C. Colmenares and A. Magdziarz, J. Mol. Catal. A: Chem., 2013, 366, 156.
28. (a) J. C. Colmenares, A. Magdziarz, K. Kurzydlowski, J. Grzonka, O. Chernyayeva and D. Lisovytskiy, Appl. Catal., B, 2013, 134–135, 136; (b) J. C. Colmenares, A. Magdziarz, O. Chernyayeva, D. Lisovytskiy, K. Kurzydłowski and J. Grzonka, ChemCatChem, 2013, 5(8), 2270.
29. H. Kominami, H. Sugahara and K. Hashimoto, Catal. Commun., 2010, 11, 426.
30. M. Morishita, Y. Shiraishi and T. Hirai, J. Phys. Chem. B, 2006, 110, 17898.
31. A. Molinari, M. Montoncello, R. Houria and A. Maldotti, Photochem. Photobiol. Sci., 2009, 8, 613.
32. A. Molinari, M. Bruni and A. Maldotti, J. Adv. Oxid. Technol., 2008, 11, 143.
33. C. Minero, A. Bedini and V. Maurino, Appl. Catal., B, 2012, 128, 135.
34. I. A. Shkrob, T. W. Marin, S. D. Chemerisov and M. D. Sevilla, J. Phys. Chem. C, 2011, 115, 4642.
35. V. Augugliaro, M. Bellardita, V. Loddo, G. Palmisano, L. Palmisano and S. Yurdakal, J. Photochem. Photobiol., C, 2012, 13, 224.
36. M. Yasuda, A. Miura, R. Yuki, Y. Nakamura, T. Shiragami, Y. Ishii and H. Yokoi, J. Photochem. Photobiol., A, 2011, 220, 195.
37. M. Tian, J. Wen, D. MacDonald, R. M. Asmussen and A. Chen, Electrochem. Commun., 2010, 12, 527.
38. A. Tanaka, K. Hashimoto and H. Kominami, J. Am. Chem. Soc., 2012, 134, 14526.
39. G. Palmisano, E. Garcia-Lopez, G. Marcì, V. Loddo, S. Yurdakal, V. Augugliaro and L. Palmisano, Chem. Commun., 2010, 46, 7074.
40. T. Ruether, A. M. Bond and W. R. Jackson, Green Chem., 2003, 5, 364.
41. Q. Wang, M. Zhang, C. Chen, W. Ma and J. Zhao, Angew. Chem., Int. Ed., 2010, 49, 7976.
42. (a) S. Higashimoto, N. Kitao, N. Yoshida, T. Sakura, M. Azuma, H. Ohue and Y. Sakata, J. Catal., 2009, 266, 279; (b) S. Higashimoto, N. Suetsugu, M. Azuma, H. Ohue and Y. Sakata, J. Catal., 2010, 274, 76; (c) S. Higashimoto, K. Okada, T. Morisugi, M. Azuma, H. Ohue, T. H. Kim, M. Matsuoka and M. Anpo, Top. Catal., 2010, 53, 578.
43. S. Kim and W. Choi, J. Phys. Chem. B, 2005, 109, 5143.
44. C.-J. Li, G.-R. Xu, B. Zhang and J. R. Gong, Appl. Catal., B, 2012, 115–116, 201.
45. L. Samiolo, M. Valigi, D. Gazzoli and R. Amadelli, Electrochim. Acta, 2010, 55, 7788.
46. H. Li, R. Liu, S. Lian, Y. Liu, H. Huang and Z. Kang, Nanoscale, 2013, 5, 3289 RSC.
47. D. Tsukamoto, Y. Shiraishi, Y. Sugano, S. Ichikawa, S. Tanaka and T. Hirai, J. Am. Chem. Soc., 2012, 134, 6309.

48. N. Zhang, S. Liu and Y.-J. Xu, Nanoscale, 2012, 4, 2227.

49. U. R. Pillai and E. Sahle-Demessie, J. Catal., 2002, 211, 434.

50. D. Spasiano, L. P. Prieto Rodriguez, J. Carbajo Olleros, S. Malato, R. Marotta and R. Andreozzi, Appl. Catal., B, 2013, 136–137, 56.

CHAPTER 12

Development of Mesoscopically Assembled Sulfated Zirconia Nanoparticles as Promising Heterogeneous and Recyclable Biodiesel Catalysts

SWAPAN K. DAS AND SHERIF A. EL-SAFTY

12.1 INTRODUCTION

The design and synthesis of porous crystalline metal oxides have been receiving great interest from the scientific community because of their versatile applications, such as solar cells, sensors, filters and catalysts. [1–7] Metal oxides fabricated via conventional sol-gel procedures are usually amorphous in nature, and their crystallization require high-temperature treatment, which often results in the collapse of mesostructures. Thus, their applications are limited. [8] This limitation has motivated researchers in studying the formation of uniform crystalline-building nanounits that can be utilized as precursor artificial atoms for the fabrication of crystalline mesoporous structure in mild reaction conditions. [9] Different routes, such as sol-gel, hydrothermal, and solvothermal crystallization, for the synthesis of metal oxide nanocrystals of various sizes and shapes and several nanoparticle (NP) assemblies with porous structures, such as TiO_2,

Development of Mesoscopically Assembled Sulfated Zirconia Nanoparticles as Promising Heterogeneous and Recyclable Biodiesel Catalysts. © Das, S. K. and El-Safty, S. A. (2013); ChemCatChem, 5: 3050–3059. doi: 10.1002/cctc.201300192. Used with permission.

[10] CeO_2, [4,11] CeO_2-$Al(OH)_3$, [11] and SnO_2, [12] have been reported. Chen et al. reported the hydrothermal fabrication of nanocrystalline zirconia (ZrO_2) by using a mixed-surfactant route with mixed tetragonal and monoclinic phases that exhibit both macropores and small mesopores.[13] We recently prepared self-assembled mesoporous ZrO_2 NPs through evaporation-induced self-assembly (EISA) method via a nonaqueous route in the presence of pluoronic F127. [14] However, the synthesis of crystalline porous self-assembled ZrO_2 NPs with controllable pore structures by using a template as a fastening agent is still a big challenge today.

The utilization of porous solid materials in biodiesel production is a very interesting area because it correlates with environmental issues and fuel crisis. [15–16] Biodiesel is considered as a promising alternative fuel compared with conventional petroleum diesel because the former is eco-friendly and promotes a positive life cycle-energy balance. Biodiesel also has a higher octane number and a higher flash point than diesel fuel, thereby providing better performance and safer use. [17–18] Moreover, alkyl esters of fatty acids have been utilized as an important ingredient in different uses, such as perfumes, flavors, cosmetics, food additives, lubricants for textiles, detergents, soaps, and plasticizers. Biodiesel is made from renewable resources and low-cost feedstock, such as yellow greases, rendered animal fats, and trap greases that contain a high concentration of free fatty acids. [19–20] Biodiesel is prepared through the esterification of free fatty acids or the transesterification of triglycerides with methanol or other alcohol in the presence of acid or base catalysts. [21–23] Industrial biodiesel production involves homogeneous base catalysis, which is correlated with a high production cost because this method encounters neutralization and separation problems. [24] Base-catalyzed procedure suffers from several limitations of feedstock. For example, the free fatty acid content in the feedstock should be lower than 0.5 wt%. Otherwise, biodiesel production will be badly hampered because of soap formation. [24] Concentrated sulfuric acid-catalyzed esterifications are basically homogeneous process, which is corrosive and critical for waste separation. [25]

Solid acid catalysts are considered as suitable candidates for heterogeneous catalysis because of the limitations in homogeneous catalysis. Heterogeneous catalysis that involves the use of solid acid catalysts does not encounter corrosion issues and offers the usual advantages such as easy product isolation and catalyst reusability, thereby minimizing the loss of products during catalyst separation. [26] To date, various solid acid catalysts such as zeolites, [27–28] sulfonated carbonized sugar, [29–30] sulfated ZrO_2, [31–33] sulfated silica-ZrO_2, [34] Zr-PMOs, [35] ion exchange resins and ionic liquid,[36] zirconium phosphate/metal oxides, [37] and organosulfonic acid-functionalized mesoporous silica [38–39] have been developed in biodiesel reaction. Zeolites are microporous solids and are not a suitable candidate in biodiesel reaction because of the diffusion limitations of long-chain fatty acid molecules. Ion exchange resins are also not considered as a potential candidate because of their low thermal stability. Solid acids usually catalyze the esterification reaction of fatty acids with methanol or with other small-chain alcohols at high temperature ranging from 373 K to 453 K. Kiss et al. have displayed various interesting biodiesel catalysts, such as niobic acid, sulfated ZrO_2, sulfated titania, and sulfated tin oxide, and have revealed that sulfated ZrO_2 is the most active in this aspect. [28,17] Another sulfated ZrO_2-anchored mesoporous silica catalyst has also been reported by Chen et al. for the heterogeneous catalysis of the esterification of long-chain fatty acids. [34] Myristic acid has been esterified with short-chain alcohols by using sulfated ZrO_2. [31] Rebeca et al. showed that Zr-loaded mesoporous organic-inorganic hybrid silica catalyzed the biodiesel production by the esterification/transesterification of free fatty acid in feedstock. [35] Thus, we have also developed zirconium oxophosphates, where the esterification reaction of different long chain fatty acids with methanol was heterogeneously catalyzed. [40] Studies have shown that researchers are extremely interested in developing heterogeneous recyclable catalysts in bio-fuel preparation, and that the invention of suitable green catalysts is a big challenge today.

In this context, we show the development of new mesoscopic assembly zirconium nanostructures (MAsZrNPs) by using premade NPs as a framework building block through the use of a template as a fastening agent in an acidic aqueous medium. The porous frameworks were generated by removing the fastening template molecules as well as the internal rearrangement of NPs at high-temperature calcinations. The highly crystalline characteristic of the NPs effectively sustained the local strain during mesophase formation. The pore walls of the materials were composed of individual NPs, which provide high structural stability in high temperature and in harsh chemical reactions. The utilization of premade NPs minimized the possibility of increased NP size during the entire porous structure creation, and provided a higher surface area compared with bulk NPs. The sulfonated matrix was used as an efficient heterogeneous solid acid catalyst in the biodiesel reaction, such as in the conversion of long-chain fatty acids to their corresponding esters in mild reaction conditions. A high surface area facilitated the integration of sulfate functionality and the open framework structure provided easy access of active site in the chemical reactions. Large pore size favors the diffusion of large-size fatty acid molecules. With the presence of strong acid sites, the biodiesel reaction rate significantly accelerated in optimal reaction conditions. [40–41] The sulfated MAsSZrNPs function as a heterogeneous and recyclable catalyst in the biodiesel reaction, and the maximum biodiesel yield was ca. 100%. To our knowledge, the fabrication of such mesoporous nanoassemblies by using premade monodisperse NPs with a high surface area and by using sodiumdodecyl sulfate (SDS) as a fastening agent via the hydrothermal method and its utilization as a heterogeneous solid acid catalyst in biodiesel reactions have not been explored. The mesopurous nanoassemblies provide efficient catalytic reusability in biodiesel reaction with negligible loss of activity.

12.2 RESULTS AND DISCUSSION

12.2.1 FABRICATION OF MASZRNPS BIODIESEL CATALYST

The development of new MAsZrNPs by using premade NPs as a framework building block through the use of a template as a fastening agent (Scheme 1) is highly challenging. The entire synthetic process involves the following: (1) preparation of highly disperse and very small ZrO_2 NPs sol and their utilization as framework building units. (2) Fastening of ZrO_2 NPs with a template molecule because the pH of the solution (pH ca. 1) is below the point of zero charge (PZC, 4-6); [14] thus, positively charged ZrO_2 NPs interact with the negative head group of the anionic structure directing agent (SDS) through electrostatic interaction, and a mesoscopic

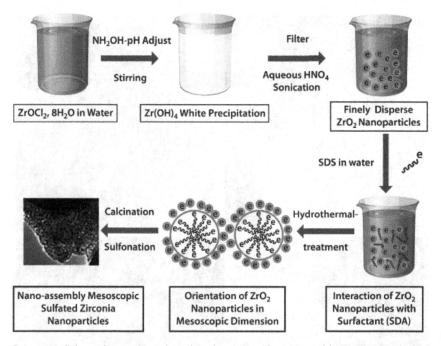

Scheme 1. Schematic representation of the formation of nanoassembly MAsZrNPs through the hydrothermal method by using an anionic structure directing agent (SDS) as a fastening agent.

assembly architecture is generated. (3) The generation of porous frameworks through the removal of fastening template molecules, the internal rearrangement of NPs at high temperature calcinations, and the highly crystalline nature of NPs effectively sustain local strain because of mesophase formation. The main advantages of these frameworks are as follows: (1) individual NP composed the pore walls of the matrix, which provide a high structural stability at high temperature and in harsh chemical reaction conditions. (2) Premade NPs minimize the possibility of NP size increase throughout the porous framework construction. (3) The special arrangement of NPs generates porous frameworks with a higher surface area compared with bulk NPs. Hence, catalytic application is important because the catalyst has plenty of accessible active sites. Moreover, large pores and an open framework structure facilitate the diffusion of large

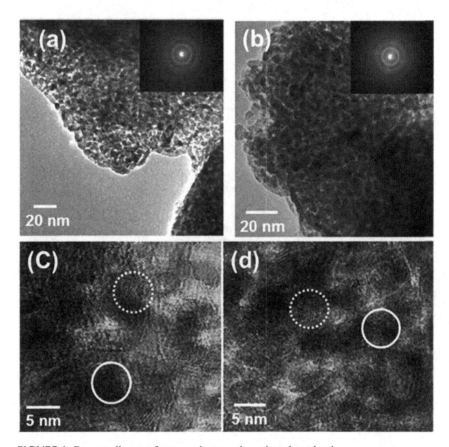

FIGURE 1: Process diagram for second generation ethanol production.

fatty acid molecules. NP morphology and crystallinity were confirmed via HRTEM, SAED (Figure 1), and wide angle powder x-ray diffraction (Figure 2B). The small-angle PXRD results of both types of materials (i.e., not sulfated and sulfated) suggest that the structure was retained and that the mesoscale was porous (Figures 2A and S1A). The EDS surface chemical analysis results show the successful integration of sulfate functionality (Figure S2, see SI). The main advantages of the materials in the catalysis are as follows: (1) high surface area that integrates sulfate functionality, and an open framework structure that provides easy access of the active

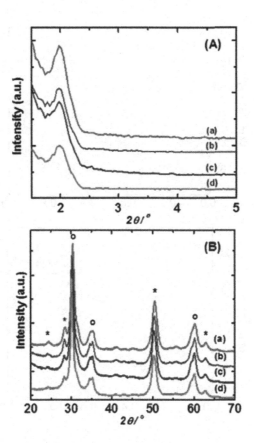

FIGURE 2: Small-angle PXRD patterns of (A) sulfated MAsSZrNPs-1 (a), MAsSZrNPs-2 (b), MAsSZrNPs-3 (c), and MAsSZrNPs-4 (d) samples; (B) Wide-angle sulfated MAsSZrNPs-1 (a), MAsSZrNPs-2 (b), MAsSZrNPs-3 (c) and MAsSZrNPs-4; (*) monoclinic phase, (o) tetragonal phase.

sites during the chemical reaction. (2) The sulfated MAsSZrNPs function as heterogeneous and recyclable catalyst in the biodiesel reaction during the transformation of long-chain free fatty acids into their corresponding esters. The maximum biodiesel yield was ca. 100%. (3) Excellent heterogeneous catalytic activity of MAsSZrNPs, which can be attributed to the large surface area, the presence of ample acidic sites located at the surface of the matrix, and its stability towards harsh chemical reaction conditions. Two NH_3 desoption peaks are assigned at 278.9 and 639 K for MAsSZrNPs (see Figure S3, in SI). Compared to the desorption peak of

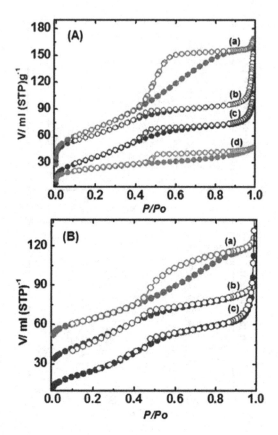

FIGURE 3: N$_2$ adsorption/desorption isotherms of (A) calcined MAsZrNPs-1 (a), MAsZrNPs-2 (b), MAsZrNPs-3 (c), and MAsZrNPs-4 (d) samples; (B) calcined mesoporous sulfated MAsSZrNPs-1 (a), MAsSZrNPs-2 (b), and MAsSZrNPs-3 (c) samples measured at 77 K. Adsorption points are marked by filled circles and desorption points by empty circles.

recently prepared sulfated zirconia at 635 K [14], the MAsSZrNPs showed a peak at much higher temperature. These findings reveal that, the fabricated strategy of nano-assembly sulfated of Zirconia showed a strongly bound of ammonia on highly acidic sites. (4) The catalyst shows negligible loss of activity in the catalytic recycles. This concept will be utilized to design and synthesize other mesoscopic assembly materials. Catalytic significance may confer its paramount importance in other acid catalyzed reactions.

12.2.1.1 HIGH-RESOLUTION TEM ANALYSIS

The representative TEM images of the calcined MAsZrNPs are shown in Figures 1a and 1b. In these images, low electron density spots (pores) are observed throughout the respective specimens, and the particles of size ca. 7.0 nm to 9.0 nm are arranged in a mesoscopic order. Interparticle pores, as seen in these images, vary from 4.0 nm to 6.0 nm in length. The SAED analysis results (Inset of Figures 1a and 1b) of the mesoporous MAsZrNPs confirm that the pore walls of our studied materials are made up of nanocrystalline oxides that show characteristic diffuse electron diffraction rings. These well-resolved diffraction rings are attributed to the polycrystalline nature of NPs. [13–14] HRTEM images reveal the orientation of NPs within the pore walls, as well as several nanocrystallites with well-resolved lattice planes, as shown in Figures 1c and 1d. In these HRTEM images, the marked white portion within the circle are the pores, whereas the marked black portion within the circle are the observed NPs. The average pore diameters measured from this micrograph are consistent with the pore size distribution derived using the NLDFT method from the N_2 adsorption/desorption isotherms, as shown in Figure 4.

12.2.1.2 POWDER X-RAY DIFFRACTION

The small-angle PXRD patterns for the calcined and sulfated MAsZrNPs are shown in Figures 2A and S1A (see supporting information; SI), respectively. Both types of materials show a single broad peak in their respective small-angle PXRD patterns. This result suggests that NPs arranged more or less disorderly in these materials, as shown in the TEM images (Figure 1). [14] The calcined MAsZrNPs show a diffraction peak at ca. 2.00° (2θ), which indicates a d spacing of ca. 4.41 nm, whereas in MAsSZrNPs, this diffraction peak slightly shifts to ca. 1.95°, and the corresponding d spacing is ca. 4.52 nm. The XRD results are consistent with the TEM micrographs and the N_2 sorption analysis results. The small-angle PRXD results show the existence of mesoporosity in both types of materials, and during high-temperature calcinations, individual NPs rearranged to create meso-

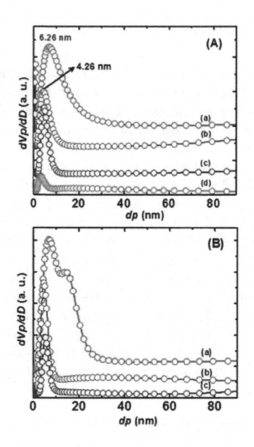

FIGURE 4: NLDFT pore size distribution of (A) calcined MAsZrNPs-1 (a), MAsZrNPs-2 (b), MAsZrNPs-3 (c), and MAsZrNPs-4 (d) samples; (B) calcined sulfated MAsSZrNPs-1 (a), MAsSZrNPs-2 (b), and MAsSZrNPs-3 (c) samples.

porosity. The highly crystalline nature of the materials effectively sustains the local strain during mesophase formation. [4,11]

The wide-angle PXRD patterns of the calcined and sulfated MAsZrNPs are shown in Figures 2B and S1B (see in SI), respectively. The XRD results for both types of materials exhibit a mixture of well-resolved characteristic monoclinic and tetragonal phases of individual ZrO_2 NPs.

[13,42–44] Calcined MAsZrNPs possesses a tetragonal phase, which is its major characteristic (Figure 2B). After sulfate integration, the mono-clinic phase becomes more prominent, as observed in the Figure S1B. Thus, integrated sulfate ions have a strong influence on phase modification. Sulfate ions converted the meta-stable tetragonal phase to its more thermodynamically stable monoclinic phase. [42] In the current research, the particle sizes of NPs were calculated using the Scherrer equation. The estimated particle size varied from 7.0 nm to 9.0 nm (see in SI). These results are in agreement with TEM image analysis results (Figure 1).

12.2.1.3 N_2 SORPTION STUDY

N_2 sorption studies are important to determine the porous nature of the materials. N_2 sorption measurements were carried out on the calcined and sulfated mesoporous MAsZrNPs at 77 K, and the type-IV adsorption-desorption isotherm exhibited a small hysteresis loop, as shown in Figures 3A and 3B, respectively. Type-IV isotherms are characteristic of meso-porous materials, and desorption hysteresis results suggest the existence of large mesopores in the sample. [45] This hysteresis is an intermediate between typical H_2- and H_4-type hysteresis loops in a P/P_0 range from 0.40 to 0.9. This result suggests that large mesopores with cage-like pore structure are connected by windows with small sizes. [10,46] This characteristic is typical of of capillary condensation within uniform pores. In the current research, the Brunauer-Emmett-Teller (BET) surface area, average pore diameter, and pore volume for the calcined and sulfated mesoporous MAsZrNPs are shown in Table 1 (entries 1–4 for the calcined MAsZrNPs and entries 4–7 for the sulfated MAsZrNPs). The BET surface area of the calcined MAsZrNPs-1, MAsZrNPs-2, MAsZrNPs-3, and MAsZrNPs-4 are 183, 142, 139, 67 m2g-1, respectively. The surface area of the sulfated MAsZrNPs-1, MAsZrNPs-2, and MAsZrNPs-3 are 93, 103, and 98 m^2g^{-1}, respectively. The pore volumes of the corresponding calcined materials decreased after the incorporation of the sulfate group, as shown in Table 1 (entries 4–7). Thus, the surface areas as well as the pore volumes of the calcined matrices decrease upon sulfate integration, which can be attributed to the dispersion of sulfate groups on the surface of the porous frame-

TABLE 1: Physico-chemical properties of mesoscopic assembly zirconia and sulfated zirconia nanoparticles.

Entry	Sample type	Surface area (m^2g^{-1})	Pore width (nm)	Pore volume (ccg^{-1})	Particle size (nm)
1	MAsZrNPs-1	183	6.26	0.272	7.63
2	MAsZrNPs-2	142	4.26	0.265	8.13
3	MAsZrNPs-3	139	4.24	0.207	8.34
4	MAsZrNPs-4	67	4.02	0.082	8.69
5	MAsSZrNPs-1	93	6.87/14.20	0.160	7.94
6	MAsSZrNPs-2	103	4.59	0.114	7.21
7	MAsSZrNPs-3	98	4.87	0.272	8.21

work. Moreover, a kind of pore blocking can also occur. [26] The pore size of MAsZrNPs, as shown in Figure 4A and estimated by employing the NLDFT method, is in agreement with the pore widths obtained from TEM images (Figure 1) and XRD analysis results (Figures 2A and 1SB).

12.2.1.4 UV-VIS DIFFUSE REFLECTANCE SPECTRA

UV-visible spectroscopy was performed to characterize the optical properties of ZrO_2 nanocrystals, where pore walls of the mesoporous self-assembly architecture were formed. Figures 5 and S3 (see SI) show the UV-visible diffuse reflectance spectra of the calcined and sulfated mesoporous MAsZrNPs, respectively. The spectral features of these materials are almost identical. The absorption band ca. 208 nm appeared because of the ligand-to-metal charge-transfer transition (LMCT, $O^{2-} \rightarrow Zr^{4+}$). Furthermore, a weak broad band is observed in the region from 224 nm to 230 nm. This result can be attributed to the presence of Zr-O-Zr linkages in the framework. [34] The UV absorption edge wavelength is very sensitive to the particle size of semiconductor nanocrystals. [47–48] For NPs (<10 nm), the band gap energy increases with decreasing crystal size, and the absorption edge of the interband transition is blue-shifted. Such blue shifts of the interband transition energy (band gap) are clearly observed

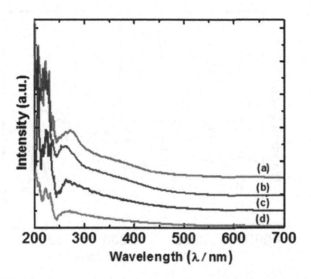

FIGURE 5: UV-Vis reflectance spectra of calcined MAsZrNPs-1 (a), MAsZrNPs-2 (b), MAsZrNPs-3 (c), and MAsZrNPs-4 (d) samples.

in the UV region of DRS for very small ZrO_2 nanocrystals because of the appearance of an additional peak in the region from 270 nm to 290 nm. [14] This spectroscopic result suggests that ZrO_2 nanocrystals compose the pore walls of the mesoporous MAsZrNPs structure. However, unlike conventional mesoporous materials with a continuous pore wall, the pore wall of MAsZrNP materials can be considered as composed of discrete nanodomains separated by voids.

12.2.1.5 FT IR SPECTROSCOPY

The FT IR spectra of the calcined and sulfated MAsSZrNPs are shown in Figure 6. The absence of bands at ~2854 and ~2925 cm⁻¹ in these samples, which are ascribed to the symmetric and asymmetric vibrations of the C-H groups, indicates the complete removal of surfactant molecules after calcinations. The broad bands at ~3000 to 3600 and 1620 cm⁻¹ are attributed to

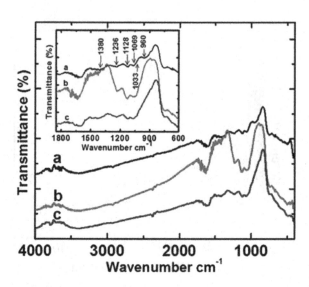

FIGURE 6: FT IR spectra A: Sulfated MAsZrNPs-1 (a), MAsZrNPs-2 (b), and MAsZrNPs-3 samples. B: In the inset is the corresponding FT IR spectra ranging from 1800 cm^{-1} to 800 cm^{-1}.

the asymmetric OH stretching and vibration bending of the adsorbed water molecule, respectively. [48] The spectral feature ranging from 1400 cm^{-1} to 900 cm^{-1} is very important in characterizing the presence of sulfate moieties in MAsSZrNPs, and all MAsSZrNP materials show almost similar spectral features. The observed band at 960 cm^{-1} indicates S-O symmetric stretching, whereas other bands at higher frequencies, such as 1033, 1069, 1128 and 1236 cm^{-1}, are attributed to S-O asymmetric stretching. [49] Furthermore, the presence of other band at 1380 cm^{-1} indicates the asymmetric stretching of S=O, which is bonded to the ZrO$_2$ NPs. [50] From the spectral data, MAsSZrNPs-2 has numerous intense bands in the sulfate region of asymmetric stretching, and this result may explain the higher catalytic efficiency of MAsSZrNPs-2 in biodiesel reaction (acid-catalyzed reaction) compared with the other two samples. Thus, this spectral investigation describes the integration of sulfate moieties into the ZrO$_2$ NPs. The partial ionic nature of the S-O bonds is responsible for the strong Brønsted acidity of the sulfate-modified ZrO$_2$ NPs. [51] The high surface area of the mesoporous MAsZrNPs facilitates the integration of sulfate functionality to a suitable extent within its framework; hence, acid-catalyzed reactions are accelerated.

12.2.1.6 CATALYTIC REACTION

The catalytic performance of sulfated MAsZrNPs in biodiesel synthesis has been investigated through the esterification of different long-chain fatty acids with methanol. In this study, methanol was used as a reactant as well as a solvent for the liquid phase. In all cases, only the methyl ester of the corresponding fatty acid was formed as a product with 100% selectivity. The results of the catalytic activities of MAsSZrNPs are given in Tables 2. In view of their good catalytic activity, a sulfate-loaded MAsSZrNPs-2 catalyst was considered, typically 2.14 wt% catalyst with respect to fatty acid (oleic acid), to obtain a maximum yield of 100% (n_{Zr}/n_S molar ratio of 16.54) with TON of 18.53 (Table 2, entry 13) at 323 K after 8 h of reaction time. When the fatty acid (oleic acid) to alcohol (methanol) molar ratio was changed to 10, the yield and TON were not significantly affected (Table 2, entry 15). TON calculation was carried out based on chemical analysis, as shown in the supporting information (Figure S2). A decrease in n_{Zr}/n_S to 15.13 (MAsSZrNPs-1), which corresponds to an increase in the integrated sulfate group amount in the matrix, lowers the yield to 77% and reduces TON to 14.26 (Table 2, entry 14). MAsSZrNPs-1 at 3 wt% results in a lauric acid methyl ester yield of 89% with TON of 17.11 (Table 2, entry 8), whereas 6 wt% of the respective catalyst results in a yield of 91%. The increase is small, but TON decreases to 8.75 (Table 2, entry 9) without changing other reaction parameters.

Temperature is a relevant factor. Thus, lauric acid was used as a reference to study the effect of temperature. For example, the temperature was lowered to 283 K in the case of MAsSZrNPs-2, which results in a decrease of the yield to 38% and TON to 7.04 (Table 2, entry 17), whereas, increasing the temperature to 303 K results in an increase of the yield to 86% and TON to 15.93 (Table 2, entry 6). Increasing the temperature of the reaction to 323 K increases the yield to 96% and TON to 17.78 (Table 2, entry 10). When the system was catalyzed only without the sulfated MAsZrNPs-1

TABLE 2: Bio-diesel reaction of different long chain fatty acids with methanol catalyzed by mesoscopic assembly zirconia and sulfated zirconia nanoparticles.

Entry	Catalyst type	Fatty acids	Alcohol: acid molar ratio	Catalyst (wt %)	Temp (K)	Yield (%)	TON[a]
1	MAsSZrNP-1	Decanoic acid	10	3.48	303	57	10.96
2	MAsSZrNP-2	Decanoic acid	10	3.48	303	86	15.93
3	MAsSZrNP-1	Decanoic acid	10	3.48	323	81	15.58
4	MAsSZrNP-2	Decanoic acid	10	3.48	323	97	17.96
5	MAsSZrNP-1	Lauric acid	10	3.00	303	76	14.61
6	MAsSZrNP-2	Lauric acid	10	3.00	303	86	15.93
7	MAsSZrNP-3	Lauric acid	10	3.00	303	42	7.78
8	MAsSZrNP-1	Lauric acid	10	3.00	323	89	17.11
9[b]	MAsSZrNP-1	Lauric acid	10	6.00	323	91	8.75
10	MAsSZrNP-2	Lauric acid	10	3.00	323	96	17.78
11[c]	MAsSZrNP-2	Lauric acid	20	3.00	303	93	17.22
12[d]	MAsSZrNP-2	Lauric acid	20	3.00	323	95	17.59
13	MAsSZrNP-2	Oleic acid	20	2.14	323	100	18.53
14	MAsSZrNP-1	Oleic acid	10	2.14	323	77	14.26
15	MAsSZrNP-2	Oleic acid	10	2.14	323	99	18.34
16	MAsSZrNP-3	Oleic acid	10	2.14	323	72	13.33
17[e]	MAsSZrNP-2	Lauric acid	10	3.00	283	38	7.04
18[f]	MAsZrNP-1	Lauric acid	10	3.00	323	21	4.32
19[g]	—	Lauric acid	10	—	323	4	—

Conditions, unless stated otherwise: The esterification of long-chain fatty acids with methanol was carried out with a mole ratio of fatty acid/methanol of 1:10 and 1:20 (entries 11–13). About 30 mg of the catalyst and 5 mmol fatty acid were used for each set of the reaction. [a]TON (turnover number) = moles of product or yield/mole of active sites (e.g., disperse Zr sites) of the catalyst. [b]Esterification of lauric acid with double amount of (60 mg) catalyst at 323 K. [c]Esterification of lauric acid with a mole ratio of fatty acid/methanol of 1:20 at 303 K. [d]Esterification of lauric acid with a mole ratio of fatty acid/methanol of 1:20 at 323 K. [e]Esterification of lauric acid at 283 K. [f]Esterification of lauric acid in the presence of un-sulfated MAsZrNPs-1 catalyst. [g]Esterification of the lauric acid in absence of catalyst. In each case, the product selectivity was 100%, and the conversion of all the products was measured after 8 h.

catalyst, the yield decreased to 21% and TON to 4.32 (Table 2, entry 18) during lauric acid esterification at 323 K for 8 h. Thus, sulfate functionalization is important for the efficient biodiesel reaction of different fatty acids with methanol in mild reaction conditions. Blank experiments confirm that fatty acid esterification does not occur considerably in the absence of MAsSZrNPs catalysts (Entry 19, Table 2). From the catalytic results, the MAsSZrNPs-2 catalyst provides the greatest efficiency in biodiesel reaction than the other two materials, and also provides the maximum yield of oleic acid.

Methanol was used in large excess relative to the fatty acids because it functions as a reactant as well as a solvent in this type of reaction. The use of a small amount of methanol (stoichiometry 1:1, e.g., methanol:fatty acid) relative to the fatty acid in the respective biodiesel reaction provides a viscous mixture of the reactant and product, which is difficult to separate. [23] Excess methanol helps dissolve the fatty acid; thus, other solvents are not needed. Methanol also facilitates the conversion to the desired product. Excess methanol used in the reaction can be reused in the next run. The boiling point of methanol is quite low; thus, the separation of excess methanol from the product mixture is not difficult. Particularly, the isolation of the biofuel product becomes much easier when methanol is used in excess in the reaction mixture.

The use of surface sulfate-modified zirconia nanoparticles is vital in the esterification of fatty acids, as shown in Figure 7. Studies have shown that sulfated ZrO_2 has a strong Brønsted acidity, and usually favors this type of reaction. [14,22,34,40,51,52] In the reaction, the long-chain fatty acid is dissolved in methanol in the presence of a catalyst (MAsSZrNPs), and the polar carboxylic group (–COOH) of the fatty acid is adsorbed to the catalytic active sites on the surface of the material. Large pores facilitate the diffusion of bulky sized fatty acid molecules through the pore interior, and a high surface area provides plenty of active sites for the reaction. The presence of sulphonic acid on the surface of the catalyst leads to a slight positive charge ($\delta+$) on the carbonyl carbon of the reacting fatty acid through the protonation of the adjacent oxygen atom. Consequently, the nucleophilic methanol (MeOH) molecule attacks the carbonyl carbon because the latter has a very high electrophilicity. After internal rearrangement, the methyl ester of the corresponding fatty acid and water are

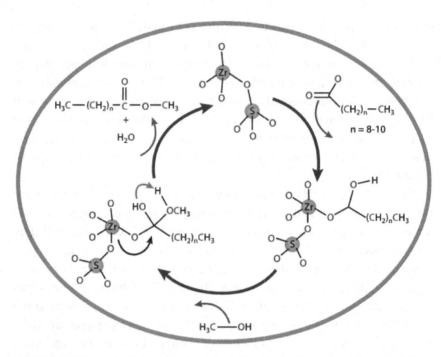

FIGURE 7: Schematic of the catalytic cycle during biodiesel reaction of long-chain fatty acids with methanol by using calcined sulfated MAsSZrNPs.

formed in the reaction. In the end, the final products are easily desorbed from the surface of the catalyst to the bulk reaction medium because the polarity of the esterified products is lower compared to that of the reactant (free fatty acids). In general, the MAsZrNPs catalysts showed a significant catalytic activity over other heterogeneous or homogeneous acid and base catalysts [17,25,28,34,40,52,53], which also require intensive design and experimental conditions in such bio-diesel catalytic reactions (Table 3).

The stability of the MAsSZrNPs catalyst as well as the heterogeneous nature of the catalysis was tested by recycling the catalyst. The hot filtration of a MAsSZrNPs-2 catalyst solution in optimized reaction conditions allows the separation of the solid catalyst, which was then reused with fresh reagents in the same reaction conditions. No loss of catalytic activity was observed. Moreover, the filtered solution was immediately used for

TABLE 3: Comparative study of the reported bio-diesel catalyst with our studied materials

Entry	Sample	Type of Catalysis	Amount (wt%)	Fatty acid: Alcohol ratio	Temp (K)	Time (h)	Conv. (%)	Ref
1	SZ	heterogeneous	1	1	403	2	40	[17]
2	SiO$_2$-SZ	heterogeneous	5	10	361	6	90	[34]
3	Porous ZrP	heterogeneous	5	10	338	24	89	[40]
4	SBA-15-SO$_3$H	heterogeneous	10	20	358	3	84	[53]
5	HZnPS-1	heterogeneous	5	10	298	24	95	[23]
6	H$_2$SO$_4$	homogeneous	1	1	403	1	96	[25, 28]
7	MAsSZrNPs-1	heterogeneous	2.14	10	323	8	77	This work
8	MAsSZrNPs-2	heterogeneous	2.14	10	323	8	99	This work

catalysis upon the addition of methanol, and the catalyst was not active in increasing fatty acid esterification. These results indicate that during the reaction, no appreciable catalyst leaching occurs, and the reaction is essentially heterogeneous. The activity of a regenerated catalyst was inspected upon separation of the solid catalyst from a reaction mixture via filtration, washed several times with methanol and anhydrous acetone, and dried in an oven at 373 K overnight. The catalyst was subsequently activated at 473 K for 4 h under air flow, and was then utilized for the above reaction. The same procedure was repeated five times with negligible loss of activity, as shown in Figure 8. The marginal loss of catalytic activity of the sulfated catalyst in the biodiesel reaction was observed after several reuse/cycles. The loss of catalytic activity was possibly due to (i) formation of water molecules formed during the esterification reaction which assisted for the deactivation of active sites. (ii) Leaching of the sulfate sites, to negligible extent, from the catalyst in the polar alcohol medium during the course of reaction. Thus, the sulfated mesoporous MAsZrNPs described herein has a great potential to be a stable and highly active recyclable solid acid catalyst in biofuel preparation.

12.3 CONCLUSIONS

In conclusion, we presented a simple and convenient synthetic method for the fabrication of MAsZrNPs by using very fine monodisperse ZrO_2 NPs through the template pathway in an acidic aqueous medium. The self-assembly process was explained by the constructive charge interactions among NPs with micellar aggregates. The self-assembly methodology, that is, nanoparticle/surfactant fastening interactions, is general and could open up a new window to other types of mesoscopic assembly of semiconductor NPs systems, as well as oxometalate cations and anions. HR TEM and N2 sorption study results reveal the formation of large mesopores through the arrangement of NPs. The integration of sulfate made this material an excellent heterogeneous biodiesel catalyst for the effective conversion of long-chain fatty acids to their methyl esters (yield ca. 100%), as shown in Table 3. The excellent heterogeneous catalytic activity and stability could be attributed to the high surface area and acidic sites located on the surface of the MAsZrNPs catalysts. Moreover, this material provides efficient reusability as a catalyst in the designed reaction with negligible loss of activity, a feature of high importance in the heterogeneous catalysis in the face of fuel crisis and environmental concern. Catalytic significance also confers its paramount importance to other acid-catalyzed reactions.

12.4 EXPERIMENTAL SECTION

12.4.1 CHARACTERIZATION TECHNIQUES

PXRD of the samples were obtained using a D8 Advance Bruker AXS diffractometer operated at 18 kW and calibrated with a standard silicon sample. An Ni-filtered Cu-Kα (λ=0.15406 nm) radiation was used. TEM images and SAED patterns were obtained using a JEOL JEM model 2100F microscope operated at 200 kV. TEM images were obtained using a CCD camera. SAED patterns were obtained using an image plate (IP) magazine. The elemental composition of the sulfated mesoporous MAsZrNPs were estimated via SEM (FESEM, JEOL model 6500). A JEOL JEM 6500 field emission scanning electron microscope (FE-SEM) attached to an EDS and operated at 20 keV was used to determine the elemental composition of the sulfated mesoporous MAsZrNPs. The nitrogen adsorption/desorption isotherms of the samples were obtained using a BELSORP36 analyzer (JP. BEL Co., Ltd.) at 77 K. Prior to the gas adsorption/desorption measurements, all the samples were degassed at 473 K for 4 h in a high-powered vacuum. The BET specific surface area was computed using the adsorption data at a relative pressure range of P/P0 from 0.05 to 0.30. The pore size distribution was derived from the adsorption isotherms by using the nonlocal density functional theory (NLDFT). The total pore volume was estimated from the amounts adsorbed at a relative pressure (P/P$_0$) of 0.99. The NH$_3$ temperature programmed desorption (NH$_3$-TPD) measurements were carried on an Autochem 2910 instrument, and a thermal conductivity detector was used for continuous monitoring of desorbed ammonia. Prior to TPD analysis, the sample was pretreated at 400°C for 1 h in a flow of ultra-pure helium gas (40 ml/min), and the sample was cooled to 100°C in the flow of ultra-pure helium gas. The pretreated sample was then saturated with 10% anhydrous ammonia gas (balance He, 60 ml/min) at 50°C for 2 h and subsequently flushed with He (60 ml/min) at 100°C for 2 h to remove the physisorbed ammonia. The heating rate of the TPD measurements, ranging from 100°C to 700°C, was 10°C/min. ^{1}H NMR experiments were carried out on a Bruker Avance DRX 600 MHz (UltraShield Plus Magnet) NMR spectrometer at an ambient temperature. The FT IR spectra of these samples were obtained using an IR Prestige-21 spectrophotometer from Shimadzu. The UV-visible diffuse reflectance spectra were obtained using a Shimadzu 3150 spectrophotometer.

12.4.2 CHEMICALS

Anionic structure-directing agent sodiumdodecyl sulfate [$CH_3(CH_2)_{11}OSO_3Na$, SDS] and different fatty acids, such as decanoic acid, lauric acid, and oleic acid, and methanol were purchased from Sigma-Aldrich. Zirconyl chloride ($ZrOCl_2.8H_2O$), ammonia (NH_3, 28%, aqueous solution), and nitric acid (HNO_3, 60%) were obtained from Wako Chemicals. Chloroform-d ($CDCl_3$, 0.05% TMS (v/v) + 99.8 atom % D) was received from Isotec (a member of Sigma-Aldrich family). All these chemicals were used without further purification.

12.4.3 SYNTHESIS

In this study, the synthesis involves two steps. The first step describes the synthesis of ZrO_2 NPs, and second step shows the fabrication of a nano-assembly mesoscopic architecture.

12.4.4 PREPARATION OF UNIFORM MONODISPERSE ZrO_2 NPs SOL

ZrO_2 NPs have been prepared by using suitable modification of previous work published elsewhere. [54] Briefly, in a typical synthesis, 3.22 g (10 mmol) of zirconyl chloride ($ZrOCl_2. 8H_2O$) was dissolved in 100 mL of distilled water. The pH of the solution was rapidly adjusted to ca. 10 by using an NH_3 solution to form hydroxide precipitates. The precipitate was filtered and thoroughly washed with excess distilled water to remove NH_3 and chloride. The precipitate was then transferred to an aqueous acidic (HNO_3) solution and was sonicated until transparent nanoparticles sol was generated. The final pH of the solution was <1, and the generated particles remained highly dispersible without sedimentation for a prolonged period.

12.4.5 PREPARATION OF MASZRNPS

MAsZrNPs were constructed by using premade ZrO_2 NPs as building blocks. In the synthetic procedure, premade ZrO_2 NPs (1 mmol) were added to 0.320 g (1.1 mmole) of the SDS solution in 80 mL of water with vigorous stirring at an ambient temperature. The solution was stirred for 2 h. This solution was then stirred further in an oil bath at 353 K for 3 h and slowly cooled to room temperature. The self-assembled nanoparticles were separated via centrifugation and dried in a vacuum. Calcination was carried out on the as-synthesized material by slowly increasing the temperature to 873 K (1 K min^{-1} ramping rate) followed by heating at 873 K for 5 h in the presence of air to obtain template-free MAsZrNPs. This sample was designated as NAsZrNPs-1.

The other 3 materials were prepared by varying the molar ratio of the precursors, such as $XZrO_2$:$YSDS$:ZH_2O. In all these cases $X = 1$, and only Y and Z were varied. The four sets of variation were $Y = 0.56$, $Z = 2224$; $Y = 0.28$, $Z = 1112$; and $Y = 0.14$ and $Z = 556$. The sample abbreviations were NAsZrNPs-2, NAsZrNPs-3, and NAsZrNPs-4, respectively.

12.4.6 PREPARATION OF MESOSCOPIC ASSEMBLY SULFATED ZIRCONIA NANOPARTICLES (MASSZRNPS)

The sulfonation of ZrO_2 NPs was carried out by treating 1.0 g of the above prepared calcined MAsZrNPs material with 15 mL of 1 N sulfuric acid aqueous solution, followed by calcinations in air at 833 K (2 K min-1 ramping rate) for 2 h.

12.4.7 CATALYTIC CONDITIONS

The catalytic biodiesel reactions were carried out in a round bottom flask fitted with a water condenser and placed in a temperature-controlled oil bath with a magnetic stirrer. In the biodiesel reaction, different long-chain fatty acids were esterified with low molecular weight alcohol, such as methanol. In a typical batch, 5 mmol fatty acid was dissolved in a suitable amount of methanol, and an amount of catalyst was added. The solution was then placed in an oil bath at a specified temperature and was stirred for 8 h. After completion of the reaction, the product was collected by separating the solid catalyst via filtration. The methanol solvent was then separated from the product mixture (filtrate) via vacuum evaporation. We varied the catalyst amount from 3.5 wt% to 5.8 wt% depending on the substrate, and we varied the temperature from 283 K to 323 K. Moreover, we also varied the molar ratios of the fatty acids and methanol from 1:10 to 1:20, respectively. Methanol was used in large excess because it is used as a reactant as well as a solvent. The product yield was calculated via NMR spectroscopy.

12.4.8 REUSE OF THE CATALYST

At the end of the reaction, the catalyst was separated from the reaction mixture via filtration. The catalyst was thoroughly washed with methanol and n-hexane or acetone for several times to remove both non-polar and polar compounds that were adsorbed on the surface and in the interior pores of the catalysts. After washing, the catalysts were activated by heating overnight at 373 K, followed by 473 K for 4 h, and then used again for the same reaction in identical reaction conditions. The whole process was repeated five times for recycling experiments. Negligible loss of the catalytic activity was observed. The reusability of the catalyst was tested using oleic acid and MAsSZrNPs-2 as the reference fatty acid and catalyst, respectively (Figure 8). The product yields for various cycles (Figure 8) were very consistent, which suggests high catalytic efficiency as well as stability of MAsSZrNPs in the biodiesel synthesis reactions.

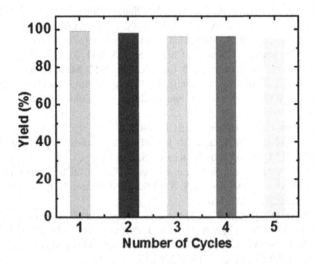

FIGURE 8: Product yields in various runs, upon catalyst recycling for biodiesel reaction of long-chain fatty acid with methanol by using sulfated MAsSZrNPs-2 catalyst. Oleic acid was used as the reference fatty acid.

REFERENCES

1. a) G. Liu, Y. Zhao, C. Sun, F. Li, G. Q. Lu, H. –M. Cheng, Angew. Chem. Int. Ed. 2008, 47, 4516-4520; b) W. Wei, C. Yu, Q. Zhao, G. Li, Y. Wan, Chem. Eur. J. 2013, 19, 566 – 577.

2. a) W. Zhou, H. Fu, ChemCatChem 2013, DOI: 10.1002/cctc.201200519; b) W. Zhou, F. Sun, K. Pan, G. Tian, B. Jiang, Z. Ren, C. Tian, H. Fu, Adv. Funct. Mater. 2011, 21, 1922-1930; c) S. A. El-Safty, M. A. Shenashen, Trends Anal. Chem. 2012, 38, 98-115.

3. a) S. A. El-Safty, M. A. Shenashen, A. A. Ismail, Chem. Commun. 2012, 48, 9652– 9654; b) S. A. El-Safty, M. A. Shenashen, M. Ismael, M. Khairy, Md. R. Awual, Analyst, 2012, 137, 5208-5214; c) S. A. El-Safty, Md.R. Awual, M.A. Shenashen, A. Shahat, Sens. Actu. B 2013, 176, 1015– 1025; d) M.A. Shenashen, A. Shahat, S. A. El-Safty, J. Hazard. Mater. 2013, 244– 245, 726– 735; e) S. A. El-Safty, M.A. Shenashen, M. Ismael, M. Khairy, Md. R. Awual, Micr. Meso. Mater. 2013, 166,195– 205; f) S. A. EL-Safty, A. Abdelllatef, M. Ismeal, A. Shahat, Adv. Healthcare Mater. 2013, DOI: 10.1002/ adhm.201200326; g) S. A. El-Safty, M. A. Shenashen, A. Shahat, Small 2013, DOI: 10.1002/smll.201202407.

4. a) S. A. El-Safty, M. Mekawy, A. Yamaguchi, A. Shahat, K. Ogawa, N. Teramae, Chem. Commun. 2010, 46, 3917-3919; b) S. A. El-Safty, A. Shahat, M. Mekawy, H. Nguyen, W. Warkocki, M. Ohnuma, Nanotechnology 2010, 21, 375603 (1-13); c) S.

A. El-Safty, A. Shahat, W. Warkocki, M. Ohnuma, Small 2011, 7, 62–65; d) S. A. El-Safty, A. Shahat, Md. R. Awual, M. Mekawy, J. Mater. Chem. 2011, 21, 5593-5603; e) S. A. El-Safty, M. A. Shenashen, Anal. Chim. Act. 2011, 694, 151-161; f) S. A. El-Safty, A. Shahat, H. Nguyen, Colloid. Surf. A 2011, 377, 44–53; g) S. A. El-Safty, N. D. Hoa, M. Shenashen, Eur. J. Inorg. Chem. 2012, 5439–5450.

5. a) A. Corma, P. Atienzar, H. Garcia, J. –Y. Chane-Ching, Nat. Mater. 2004, 3, 394-397; b) T. Waitz, T. Wagner, T. Sauerwald, C. –D. Kohl, M. Tiemann, Adv. Funct. Mater. 2009, 19, 653-661.

6. W. Schmidt, ChemCatChem 2009, 1, 53-67.

7. a) S. A. El-Safty, M. Khairy, M. Ismael, H. Kawarada, Appl. Catal. B: Environ. 2012 123-124, 162-173; b) M. Khairy, S. A. El-Safty, Curr. Catal. 2013, 2, 17-26; c) S. Das, A. Goswami, N. Murali, T. Asefa, ChemCatChem 2013, DOI: 10.1002/cctc.201200551; d) M. Khairy, S. A. El-Safty, M. Ismael, H. Kawarada, Appl. Catal. B: Environ. 2012, 127, 1-10.

8. a) P. Yang, D. Zhao, D. I. Margolese, B. F. Chmelka, G. D. Stucky, Chem. Mater. 1999, 11, 2813-2826; b) S.A. El-Safty, J. Porous Mater. 2011, 18, 259–287; c) S. A. El-Safty, A. Shahat, K. Ogawa, T. Hanaoka, Micro. Meso. Mater. 2011, 138, 51–62; d) S.A. El-Safty, J. Porous Mater. 2008,15, 369-387.

9. a) D. Wang, T. Xie, Q. Peng, Y. Li, J. Am. Chem. Soc. 2008, 130, 4016-4022; b) M. Khairy, S. A. El-Safty, M. Ismael, Chem. Commun. 2012, 48, 10832 – 10834.

10. S. K. Das, M. K. Bhunia, A. Bhaumik, Dalton Trans. 2010, 39, 4382-4390.

11. J. -Y. Chane-Ching, F. Cobo, D. Aubert, H. G. Harvey, M. Airiau, A. Corma, Chem. Eur. J. 2005, 11, 979-987.

12. a) J. Ba, J. Polleux, M. Antonietti, M. Niederberger, Adv. Mater. 2005, 17, 2509-2512; b) C. Aprile, L. Teruel, M. Alvaro, H. Garcia, J. Am. Chem. Soc. 2009, 131, 1342-1343.

13. H. Chen, J. Gu, J. Shi, Z. Liu, J. Gao, M. Ruan, D. Yan, Adv. Mater. 2005, 17, 2010-2014.

14. S. K. Das, M. K. Bhunia, A. K. Sinha, A. Bhaumik, J. Phys. Chem. C 2009, 113, 8918-8923.

15. a) Y. S. Tao, H. Kanoh, L. Abrams, K. Kaneko, Chem. Rev. 2006, 106, 896-910; b) "Statistics. The EU biodiesel industry". European Biodiesel Board., 2008-03-28. Retrieved 2008-04-03.

16. a) T. –W. Kim, M. –J. Kim, F. Kleitz, M. M. Nair, R. Guillet-Nicolas, K. –E. Jeong, H. –J. Chae, C. –U. Kim, S. –Y. Jeong, ChemCatChem 2012, 4, 687-697; b) G. Corro, U. Pal, N. Tellez, Appl. Catal. B: Environ. 2013, 129, 39-47.

17. A. A. Kiss, A. C. Dimian, G. Rothenberg, Energy & Fuels 2008, 22, 598-604.

18. a) J. Sheehan, V. Camobreco, J. Duffield, M. Graboski, H. Shapouri, National Renewable Energy Laboratory: Golden, CO, 1998, 60; b) A. Demirbas, Energy Explor. Exploit. 2003, 21, 475–487.

19. a) W. Li, S. Choi, J. H. Drese, M. Hornbostel, G. Krishnan, P. M. Eisenberger, C. W. Jones, ChemSusChem 2010, 3, 899-903; b) U.S. Dept. of Energy. Clean Cities Alternative Fuel Price Report July 2009. Retrieved 9-05-2009

20. a) R. E. Teixeira, Green Chem. 2012, 14, 419-427; b) K. L. Kline, G. A. Oladosu, A. K. Wolfe, R. D. Perlack, V. H. Dale, M. McMahon, Biofuel feedstock assess-

ment for selected countries, February, 2008, Report No. ORNL/TM-2007/224, DOI; 10.2172/924080

21. H. Fukuda, A. Kondo, H. Noda, J. Biosci. Bioeng. 2001, 92, 405-416.
22. N. Pal, M. Paul, A. Bhaumik, J. Solid State Chem. 2011, 184, 1805-1812.
23. M. Pramanik, M. Nandi, H. Uyama, A. Bhaumik, Green Chem. 2012, 14, 2273-2281.
24. E. Lotero, Y. Liu, D. E. Lopez, K. Suwannakarn, D. A. Bruce, J. G. Goodwin, Ind. Eng. Chem. Res. 2005, 44, 5353-5363.
25. a) J. M. Marchetti, A. F. Errazu, Biomass Bioenergy 2008, 32, 892-895; b) A. Macarioa, G. Giordanoa, B. Onida, D. Cocina, A. Tagarelli, A. M. Giuffrè, Appl. Catal. A: Gen. 2010, 378, 160-168.
26. M. K. Bhunia, S. K. Das, P. Pachfule, R. Banerjee, A. Bhaumik, Dalton Trans. 2012, 41, 1304-1311.
27. J. C. Juan, J. Zhang, M. A. Yarmo, Catal. Lett. 2008, 126, 319-324.
28. A. A. Kiss, A. C. Dimian, G. Rothenberg, Adv. Synth. Catal. 2006, 348, 75-81.
29. A. C. Dimian, Z. W. Srokol, M. C. Mittelmeijer-Hazeleger, G. Rothenberg, Top. Catal. 2010, 53, 1197-1201.
30. a) R. Liu, X. Wang, X. Zhao, and P. Feng, Carbon 2008, 46, 1664-1669; b) R. Luque, J. H. Clark, ChemCatChem 2011, 3, 594-597.
31. K. Saravanan, B. Tyagi, H. C. Bajaj, Catal. Sci. Technol. 2012, 2, 2512-2520.
32. a) A. Patel. V. Brahmkhatri, N. Singh, Renewable Energy 2013, 51, 227-233; b) W. Li, F. Ma, F. Su, L. Ma, S. Zhang, Y. Guo, ChemCatChem 2012, 4, 1798-1907.
33. Y. Sun, S. Ma, Y. Du, L. Yuan, S. Wang, J. Yang, F. Deng, F. –S. Xiao, J. Phys. Chem. B 2005, 109, 2567-2572.
34. X.-R. Chen, Y.-H. Ju, C.-Y Mou, J. Phys. Chem. C 2007, 111, 18731-18737.
35. R. Sánchez-Vázquez, C. Pirez, J. Iglesias, K. Wilson, A. F. Lee, J. A. Melero, ChemCatChem 2013, DOI: 10.1002/cctc.201200527.
36. a) S. Blagov, S. Parada, O. Bailer, P. Moritz, D. Lam, R. Weinand, H. Hasse, Chem. Eng. Sci. 2006, 61, 753-765; b) D. Fang, J. Yang, C. Jiao, ACS Catal. 2011, 1, 42-47.
37. K. N. Rao, A. Sridhar, A. F. Lee, S. J. Tavener, N. A. Young, K. Wilson, Green Chem. 2006, 8, 790-797.
38. a) M. I. Lòpez, D. Esquivel, C. Jiménez-Sanchidrián, F. J. Romero-Salguero, ChemCatChem 2013, DOI: 10.1002/cctc.201200509; b) T. Yalçinyuva, H. Deligöz, I. Boz, M. A. Gürkaynak, Int. J. Chem. Kinet. 2008, 40, 136-144.
39. Y. F. Feng, X. Y. Yang, D. Yang, Y. C. Du, Y. L. Zhang, F. S. Xiao, J. Phys. Chem. B 2006, 110, 14142-14147.
40. S. K. Das, M. K. Bhunia, A. K. Sinha, A. Bhaumik, ACS Catal. 2011, 1, 493-501.
41. S. K. Das, M. K. Bhunia, A. Bhaumik, The Open Catal. J. 2012, 5, 56-65.
42. B. M. Reddy, P. M. Sreekanth, P. Lakshmanan, J. Mol. Catal. A: Chem. 2005, 237, 93-100.
43. J. R. Sohn, T. –D. Kwon, S. –B. Kim, Bull. Korean Chem. Soc. 2001, 22, 1309-1315.
44. Y. –S. Hsu, Y. –L. Wang, A. –N. Ko, J. Chin. Chem. Soc. 2009, 56, 314-322.
45. S. A. El-Safty, M. A. Shenashen, M. Ismael, M. Khairy, Adv. Funct. Mater. 2012, 22, 3013-3021.
46. W. Yue, A. H. Hill, A. Harrison, W. Zhou, Chem. Commun. 2007, 2518-2520.

47. a) L. Kundakovic, M. Flytzani-Stephanopoulos, Appl. Catal. A: Gen. 1999, 183, 35-51; b) A. Bensalem, J. C. Muller, and F. Bozon-Verduraz, J. Chem. Soc., Faraday Trans. 1992, 88, 153-154.
48. S. K. Das, M. K. Bhunia, M. M. Seikh, S. Dutta, A. Bhaumik, Dalton Trans. 2011, 40, 2932-2939.
49. J. R. Sohn, S. H. Lee, J. S. Lim, Catal. Today 2006, 116, 143-150.
50. B. A. Morrow, R. A. McFarlane, M. Lion, J. C. Lavalley, J. Catal. 1987, 107, 232-239.
51. T. Yamaguchi, T. Jin, K. Tanabe, J. Phy. Chem.1986, 90, 3148-3152.
52. a) B. M. Reddy, M. K. Patil, Chem. Rev. 2009, 109, 2185-2206; b) M. K. Lam, K. T. Lee, A. R. Mohamed, Biotechnol. Adv. 2010, 28, 500-518.
53. Y. F. Feng, X. Y. Yang, D. Yang, Y. C. Du, Y. L. Zhang, F. S. Xiao, J. Phys. Chem. B 2006, 110, 14142-14147.
54. A. S. Deshpande, N. Pinna, P. Beato, M. Antonietti, M. Niederberger, Chem. Mater. 2004, 16, 2599-2604.

There are several supplemental files that are not available in this version of the article. To view this additional information, please use the citation on the first page of this chapter.

Kinetic Study on the $Cs_X H_{3-X}$ $PW_{12}O_{40}$/Fe-SiO$_2$ Nanocatalyst for Biodiesel Production

MOSTAFA FEYZI, LEILA NOROUZI, AND HAMID REZA RAFIE

13.1 INTRODUCTION

Biodiesel is a fuel composed of monoalkyl esters of long chain fatty acids derived from renewable sources, such as vegetable oils and animal fats [1]. Due to oxygen content, biodiesel is a clean, nontoxic, and biodegradable fuel with low exhaust emissions, without the sulphur and carcinogen content [2]. Biodiesel is prepared via reaction between triglycerides and alcohol in the presence of a catalyst [3]. Transesterification reaction of oil and alcohol with homogeneous catalyst is the most common method for the preparation of biodiesel [4, 5]. However, the homogeneous catalyst has many drawbacks, such as the difficulty in product separation, and equipment corrosion requirement of large quantity of water, environmental pollution [6]. The use of heterogeneous catalysts to replace homogeneous ones is easily regenerated and has a less corrosive nature, leading to safer, cheaper, and more environment-friendly operations, [7]. Currently, the research is focused on sustainable solid acid catalysts for transesteri-

Kinetic Study on the CsXH3−X PW12O40/Fe-SiO2 Nanocatalyst for Biodiesel Production. © 2013 Mostafa Feyzi et al. The Scientific World Journal, Volume 2013 (2013), Article ID 612712 (doi:10.1155/2013/612712). Creative Commons Attribution License (http://creativecommons.org/licenses/by/3.0/).

fication reaction. In addition, it is believed that solid acid catalysts have the strong potential than liquid acid catalyst [8]. HPW catalyst is a strong Brønsted acid, with high thermal stability and high solubility in polar solvents, and it has shown to be more active in FFA esterification reactions than mineral acid catalysts [8–12]. However, heteropolyacids have low surface areas and high solubility in polar solvents. These features make some difficulties in catalyst recovery and catalyst lifetime [13]. Therefore, salts of HPAs with large single valence ions, such as Cs^+, , and Ag^+, have attracted much fondness because these salts will increase in surface area and profound changes in solubility over the parent heteropolyacid [14, 15]. HPW can be supported on several kinds of support such as SiO_2, Al, ZrO_2, activated carbon, SiO_2-Al, and MCM-41 that SiO_2 is cheap, easily available, and easily surface modifiable [16–18].

In the present kinetic study, pseudo-first-order model was applied to correlate the experimental kinetic data of catalytic performance of the $Cs_xH_{3-x}PW_{12}O_{40}/Fe$-$SiO_2$ for sunflower oil transesterification with methanol and the main thermodynamic parameters such as the effects of temperature on reaction rate, activation energy, entropy variation (ΔS), and enthalpy variation (ΔH) were determined; moreover, the effects of temperature on reaction rate and the order of reaction were assessed. Characterization of catalyst was carried out by using scanning electron SEM, XRD, FT-IR, N_2 adsorption-desorption measurements methods, TGA, and DSC methods.

13.2 EXPERIMENTAL

13.2.1 FE-SIO_2 SUPPORT PREPARATION

All materials with analytical purity were purchased from Merck and used without further purification. Ferric nitrate ($Fe(NO_3)_3 \cdot 9H_2O$) and tetraethyl orthosilicate (TEOS) were selected as the source of ferric and silica, respectively. A typical procedure for the preparation of ferric-silica mixed oxide containing 60 wt% ferric was followed. Firstly, 30 mL TEOS was mixed with certain amount anhydrous ethyl alcohol (C_2H_5OH). Secondly,

35.234 gr Fe(NO$_3$)$_3$·9H$_2$O was dissolved with certain amount of anhydrous ethyl alcohol under stirring; also 30 gr of oxalic acid was dissolved in certain amount of anhydrous alcohol under stirring. In the final step, Fe and Si sols and oxalic acid were added simultaneously into the beaker under constant stirring to obtain a gel form. After the end of the above operations, the samples were aged for 90 min at 50°C. The obtained gel was dried in an oven (120°C, 12 h) to give a material denoted as the catalyst precursor. Finally, the catalyst precursor was calcined at 600°C for 6 h to produce magnetic solid catalyst. The Fe-SiO$_2$ supported 12-tungstophosphoric acid catalyst with 4 wt.% of aqueous solution HPW (based on the Fe-SiO$_2$ weight) were prepared by incipient wetness impregnation. The impregnated precursor was dried at 120°C for overnight and calcined at 600°C for 6 h.

Finally, the promoted catalyst by Cs was synthesized with Cs/H$_3$PW$_{12}$O$_{40}$ = 2 wt.%. The salt obtained by this procedure will be designated hereinafter as $Cs_xH_{3-x}PW_{12}O_{40}$/Fe-SiO$_2$, where is the amount of protons per $[PW_{12}O_{40}]^{3-}$ anion in the salt.

13.2.2 CHARACTERIZATION OF CATALYST

13.2.2.1 N$_2$ PHYSISORPTION MEASUREMENTS

The specific surface area, total pore volume, and the mean pore diameter were measured using a N$_2$ adsorption-desorption isotherm at liquid nitrogen temperature (−196°C), using a NOVA 2200 instrument (Quantachrome, USA). Prior to the adsorption-desorption measurements, all the samples were degassed at 110°C in a N$_2$ flow for 2 h to remove the moisture and other adsorbates.

13.2.2.2 SCANNING ELECTRON MICROSCOPY (SEM)

The morphology of catalyst and precursor was observed by means of an S-360 Oxford Eng scanning electron microscopy.

13.2.2.3 FOURIER TRANSFORM-INFRARED SPECTROSCOPY (FT-IR)

Fourier transform infrared (FT-IR) spectra of the samples were obtained using a Bruker Vector 22 spectrometer in the region of 400–4000 cm^{-1}.

13.2.2.4 X-RAY DIFFRACTION (XRD)

X-ray diffraction (XRD) patterns of the catalysts were recorded on a diffractometer using CuK$_\alpha$ radiation. The intensity data were collected over a 2θ range of 15–75.

13.2.2.5 THERMAL GRAVIMETRIC ANALYSIS (TGA) AND DIFFERENTIAL SCANNING CALORIMETRY (DSC)

The TGA and DSC were carried out using simultaneous thermal analyzer (PerkinElmer) under a flow of dry air with a flow rate of 50 mL min^{-1}. The temperature was raised from 20 to 700°C using a linear programmer at a heating rate of 3°C min^{-1}.

13.2.3 CATALYTIC TESTS

The type and quantity of methyl esters in the biodiesel samples were determined using gas chromatography-mass spectrometry (GC Agilent 6890N model and Mass Agilent 5973N model) equipped with a flame ionized detector (FID). A capillary column (HP-5) with column length (60 m), inner diameter (0.25 mm), and 0.25 μm film thickness was used with helium as the carrier gas. The temperature program for the biodiesel samples started at 50°C and ramped to 150°C at 10°C min^{-1}. The temperature was held at 150°C for 15 min and ramped to 280°C at 5°C min^{-1}. The holding time at the final temperature (250°C) was 5 min. Also, the injector was used from kind split/splitless.

TABLE 1: Concentration of methyl ester (based on mole fraction) at temperatures of 60 (A), 55 (B), and 50°C (C), respectively.

Reaction time (s)	X_{ME}			$-\ln(1-X_{ME})$		
	A	B	C	A	B	C
0	0.00	0.00	0.00	0.000000	0.000000	0.000000
30	0.32	0.24	0.14	0.385660	0.274437	0.150823
60	0.40	0.28	0.15	0.510826	0.328504	0.162519
90	0.42	0.34	0.24	0.590000	0.415515	0.274437
120	0.5819	0.43	0.30	0.872035	0.562119	0.356675
150	0.63	0.51	0.34	0.994252	0.713350	0.415515
180	0.70	0.57	0.38	1.203973	0.843970	0.478036
210	0.75	0.63	0.45	1.386294	0.994252	0.597837
240	0.81	0.65	0.50	1.660731	1.049822	0.693147

X_{ME}: concentration of methyl ester.
$-\ln(1-X_{ME})$: triglyceride concentration.

13.2.4 KINETIC STUDIES

The transesterification of sunflower oil was carried out in a round bottomed flask fitted with a condenser and magnetic stirring system. The reaction system was heated to selected temperature in 50, 55, and 60°C. When the oil reached selected temperature, methanol/oil molar ratio = 12/1 and the catalyst amount (3 wt% related to oil weight) were added with continuous stirring (500 rpm). After completion of the reaction time (0–240 min), the sample concentration is calculated according to mole fraction at any time. The results can be seen in Table 1.

13.3 KINETIC MODEL

The mole fraction for the transesterification reaction was established. For the reaction stoichiometry requires 3 Mol of methanol (M) and 1 mol of triglyceride (TG) to give 3 mol of methyl ester (ME) and 1 Mol of glycerol

(GL). Transesterification reaction comprises three consecutive reversible reactions, where 1 mol of ME is produced in each step and monoglycerides (MG) and diglycerides (DG) are intermediate products [19]. The kinetic model used in this work is based on the following assumptions.

(1) Firstly, k_{eq} should be considered not to depend on methanol concentration (due to its excess) (the reaction is considered pseudo-first order [20, 21]).

(2) Production of intermediate species is negligible (the result reaction is a one step).

(3) The chemical reaction occurred in the oil phase.

Based on assumption ,

$$-r = \frac{-d[TG]}{dt} = k[TG][ROH]^3 \qquad (1)$$

And based on assumption ,

$$k' = k[ROH]^3$$

$$-r = \frac{-d[TG]}{dt} = k[TG][ROH]^3 \qquad (2)$$

$$\rightarrow lnTG_0 - lnTG = k' \cdot t$$

According to the mass balance,

$$X_{ME} = 1 - \frac{[TG]}{[TG_0]}$$

$$[TG] = [TG_0][1 - X_{ME}]$$

$$\frac{dX_{ME}}{dt} = k'[1 - X_{ME}] \rightarrow -\ln(1 - X_{ME}) = k' \cdot t \qquad (3)$$

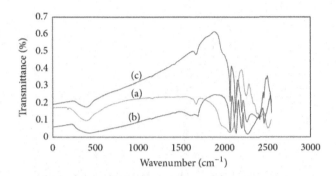

FIGURE 1: FT-IR spectrums of the $Cs_xH_{3-x}PW_{12}O_{40}$/Fe-SiO$_2$ (a), $H_3PW_{12}O_{40}$ (b), and $Cs_xH_{3-x}PW_{12}O_{40}$ (c) nanocatalysts.

Based on this model and the experimental data, at first we calculated the concentration of methyl ester at different times (based on the moles fraction). Second, the rate constants at each temperature were obtained. Third, the preexponential factors and activation energies are obtained by plotting the logarithm of the rate constants (k) versus 1/T of absolute temperature using the Arrhenius equation and in the final stage thermodynamic parameters were obtained such as ΔS and ΔH.

13.4 RESULTS AND DISCUSSION

13.4.1 CHARACTERIZATION OF THE CATALYST

The FT-IR spectra of $H_3PW_{12}O_{40}$, $PW_{12}O_{40}$ and $Cs_xH_{3-x}PW_{12}O_{40}$/Fe-SiO$_2$ are shown in Figure 1. The Keggin anion of HPW consists of a central phosphorous atom tetrahedrally coordinated by four oxygen atoms and surrounded by twelve octahedral WO6 units that share edges and corners in the structure.bands related to a Keggin structure (i(O–P–O) = 550 cm^{-1}, indicative of the bending of the central oxygen of P–O–P; vas(W–Oe–W) at 798 cm^{-1}, related to asymmetric stretching of tungsten with edge oxygen in W–O–W; vas(W–OcW) = 893 cm^{-1}, related to the asymmetric stretching of corner oxygen in W–O–W; vas(W=O) = 983 cm^{-1}, indicative of the

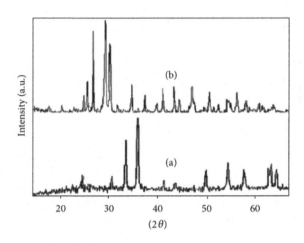

FIGURE 2: XRD patterns of the $Cs_1H_2PW_{12}O_{40}/Fe$-SiO_2 (a) and $Cs_1H_2PW_{12}O_{40}$ (b) nanocatalysts.

asymmetric stretching of the terminal oxygen; and $v(P–O)$ at $1080\,cm^{-1}$, assigned to asymmetric stretching of oxygen with a central phosphorous atom) [22, 23]. HPA salts maintain their corner Keggin structure with the addition of different amount of metallic Cs. When $Cs_xH_{3-x}PW_{12}O_{40}$ is supported on Fe-SiO$_2$, these bands have somewhat changed. The bands at 1080 and $890\,cm^{-1}$ are overlapped by the characteristic band of SiO$_2$, while these bands at 985 and $794\,cm^{-1}$ shift to 966 and $805\,cm^{-1}$, respectively.

The XRD pattern of $Cs_xH_{3-x}PW_{12}O_{40}/Fe$-$SiO_2$ catalyst was presented in Figure 2. Supported HPA samples do not show diffraction patterns probably due to the following reasons: (i) after treatment at 500°C, the HPA practically loses its crystalline structure, (ii) HPA species were highly dispersed, and (iii) the deposited amount of HPA was not big enough to be detected by this technique [24].

The TGA-DSC experiment on the catalyst precursor has shown four steps of mass loss (Figure 3). The first step at the temperature of 70–110°C was attributed to the evaporation of residual moistures in the catalyst precursor and loss of physisorbed waters. The second stage (190–280°C) is accompanying weight loss of the crystallization water expelling which is most likely the hydrated proton. The peak around 390–460°C is due to the

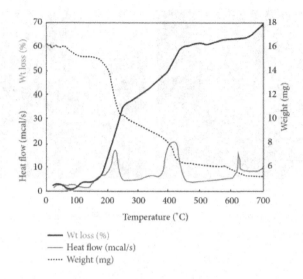

FIGURE 3: TGA and DSC curves for $Cs_xH_{3-x}PW_{12}O_{40}$/Fe-SiO_2 precursor.

decomposition of iron and Si oxalates to oxide phases. Most of weight loss happened from 580°C to 640°C due to the phase transition and formation of Fe_2SiO_4 (cubic). The weight loss curve is involved with a total overall weight loss of ca. 69 wt%. DSC measurement was performed in order to provide further evidence for the presence of the various species and evaluate their thermal behavior. As shown in Figure 3, the endothermic curve represents the removal of the physically adsorbed water from the material (70–110°C). Two exothermic peaks at around 190–280°C and 390–480°C are due to the crystallization water expelling which is most likely the hydrated proton and the decomposition of iron and Si oxalates to oxide phases, respectively. The exothermic peak at around 580–640°C is due to formation of iron silicate phase [25].

Figure 4 shows SEM pictures of the precursor (a) and calcined catalyst (b). After calcination catalyst particle aggregated together and formed a spherical shape and more uniform particles which are beneficial to the activity and augmenting the surface of the catalyst that exhibited a large amount of aggregates than the precursor. The measured BET surface areas are 237.5 $m^2 cm^3 g^{-1}$ for $Cs_xH_{3-x}PW_{12}O_{40}$/Fe-SiO_2 catalyst and corresponded pore volume is 0.7672 $cm^3 g^{-1}$ obtained from analysis of the

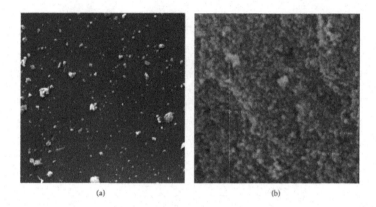

(a)   (b)

FIGURE 4: The SEM images of $Cs_xH_{3-x}PW_{12}O_{40}/Fe$-$SiO_2$ nanocatalyst, (a) precursor and (b) calcined catalyst.

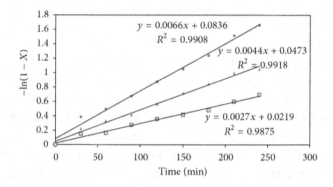

FIGURE 5: Plots of $-\ln(1-X)$ versus time (min) at temperatures 60, 55, and 50°C for reaction of sunflower oil with methanol.

desorption using the BJH (Barrett-Joyner-Halenda) method. The particle size could be calculated by Scherer-equation form XRD pattern (Figure 2). It is clear that the catalyst particle size was in nanodimension (45 nm). The $Cs_xH_{3-x}PW_{12}O_{40}/Fe$-$SiO_2$ catalyst was characterized with SEM (Figure 4). It is obvious in this figure that the crystal sizes were from 38–47 nm.

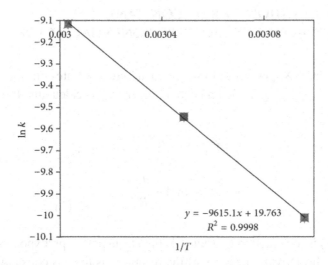

FIGURE 6: Arrhenius plot of ln k versus 1/T for reaction of sunflower oil with methanol.

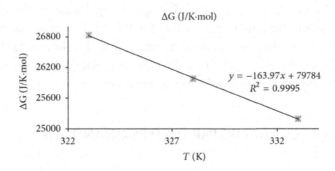

FIGURE 7: Plot of ΔG(J/K mol) versus T(K) for reaction of sunflower oil with methanol.

This result confirmed the obtained results studied by using the Scherrer equation.

13.2 CALCULATION OF RATE CONSTANT, ACTIVATION ENERGY, AND PREEXPONENTIAL FACTOR

Plots of $-\ln(1-X_{ME})$ versus (T) are given in Figure 5. Rate constant has been calculated using Figure 5. And activation energy is calculated through the Arrhenius equation:

$$k = A_e^{-E_a/RT} \qquad (4)$$

$$\rightarrow \ln k = \ln A - \frac{E_a}{RT} \qquad (5)$$

where (k) is the reaction constant, is the frequency or preexponential factor, E_a is the activation energy of the reaction, is the gas constant, and is the absolute temperature.

Therefore, plots of versus are given in Figure 6. Activation energy () and preexponential factor have been calculated using Figure 6 to be 79.805 kJ/mol and 8.9×10^8 kJ/mol, respectively. Based on the proposed kinetic model, the kinetic parameters for this catalyst were determined. The experiments demonstrate that the reactions follow first-order kinetics. The proposed kinetic model describes the experimental results well and the rate constants follow the Arrhenius equation.

13.3 THERMODYNAMIC PARAMETERS (ΔS AND ΔH)

Based on the definition of Gibbs energy free using (7) and using linear plot of $\ln k_{eq}$ versus $1/T$ which is given in Figure 7 and by using (7) the respective values of ΔS and ΔH were calculated which are 0.0197 kJ/mol, 79.784 kJ/Kmol respectively:

$$\Delta G = RT \ln k_{eq} \qquad (6)$$

$$\ln k_{eq} = \frac{\Delta S}{R} - \frac{\Delta H}{RT}$$

(7)

From the results (Table 2) of the thermodynamic parameters it can be found that the transesterification reaction is an endothermic reaction and, with increasing temperature, reaction rate increases. Moreover, enthalpy and entropy change are not affected by methanol concentration due to its excess [26].

TABLE 2: Calculated values of thermodynamic parameters.

T (K)	ΔG (kJ/K·mol)	ΔH (kJ/K·mol)	ΔS (KJ/mol)
323	26.83272		
328	25.98118	79.784	0.0197
333	25.19303		

13.5 CONCLUSIONS

The magnetic $Cs_xH_{3-x}PW_{12}O_{40}$/Fe-SiO$_2$ nanocatalyst was prepared for biodiesel production. Experimental conditions were varied as follows: reaction temperature 328–338 K, methanol/oil molar ratio = 12/1, and the reaction time 0–240 min. Thermodynamic properties such as   and   were successfully determined from equilibrium constants measured at different temperature. Activation energy (E_a) and preexponential factor have been calculated to be 79.805 kJ/mol and 8.9×10^8 kJ/mol, respectively. The experiments demonstrate that the reactions follow first-order kinetics. Also notably, recovery of the catalyst can be achieved easily with the help of an external magnet in a very short time (<20 seconds) with no need for expensive ultracentrifugation.

REFERENCES

1. M. C. Math, S. P. Kumar, and S. V. Chetty, "Technologies for biodiesel production from used cooking oil: a review," Energy for Sustainable Development, vol. 14, no. 4, pp. 339–345, 2010.

2. P.-L. Boey, G. P. Maniam, and S. A. Hamid, "Performance of calcium oxide as a heterogeneous catalyst in biodiesel production: a review," Chemical Engineering Journal, vol. 168, no. 1, pp. 15–22, 2011.

3. E. H. Pryde, "Vegetable oils as fuel alternatives: symposium overview," Journal of the American Oil Chemists Society, vol. 61, no. 10, pp. 1609–1610, 1984.

4. J. M. Dias, M. C. M. Alvim-Ferraz, and M. F. Almeida, "Comparison of the performance of different homogeneous alkali catalysts during transesterification of waste and virgin oils and evaluation of biodiesel quality," Fuel, vol. 87, no. 17-18, pp. 3572–3578, 2008.

5. Z. Helwani, M. R. Othman, N. Aziz, J. Kim, and W. J. N. Fernando, "Solid heterogeneous catalysts for transesterification of triglycerides with methanol: a review," Applied Catalysis A, vol. 363, no. 1-2, pp. 1–10, 2009.

6. L. C. Meher, D. Vidya Sagar, and S. N. Naik, "Technical aspects of biodiesel production by transesterification: a review," Renewable and Sustainable Energy Reviews, vol. 10, no. 3, pp. 248–268, 2006.

7. F. Qiu, Y. Li, D. Yang, X. Li, and P. Sun, "The use of heterogeneous catalysts to replace homogeneous ones can be expected to eliminate the problems associated with homogeneous catalysts," Bioresource Technology, vol. 102, pp. 4150–4156, 2011.

8. K. Jacobson, R. Gopinath, L. C. Meher, and A. K. Dalai, "Solid acid catalyzed biodiesel production from waste cooking oil," Applied Catalysis B, vol. 85, no. 1-2, pp. 86–91, 2008.

9. M. Zabeti, W. M. A. Wan Daud, and M. K. Aroua, "Activity of solid catalysts for biodiesel production: a review," Fuel Processing Technology, vol. 90, no. 6, pp. 770–777, 2009.

10. S. Shanmugam, B. Viswanathan, and T. K. Varadarajan, "Esterification by solid acid catalysts: a comparison," Journal of Molecular Catalysis A, vol. 223, no. 1-2, pp. 143–147, 2004.

11. Y. Liu, E. Lotero, and J. G. Goodwin Jr., "A comparison of the esterification of acetic acid with methanol using heterogeneous versus homogeneous acid catalysis," Journal of Catalysis, vol. 242, no. 2, pp. 278–286, 2006.

12. Z. Helwani, M. R. Othman, N. Aziz, W. J. N. Fernando, and J. Kim, "Technologies for production of biodiesel focusing on green catalytic techniques: a review," Fuel Processing Technology, vol. 90, no. 12, pp. 1502–1514, 2009.

13. H.-J. Kim, Y.-G. Shul, and H. Han, "Synthesis of heteropolyacid (H3PW12O40)/ SiO2 nanoparticles and their catalytic properties," Applied Catalysis A, vol. 299, no. 1-2, pp. 46–51, 2006.

14. N. Mizuno and M. Misono, "Pore structure and surface area of Cs (x = 0–3, M = W, Mo)," Chemistry Letters, vol. 16, pp. 967–970, 1987.

15. T. Okuhara, T. Ichiki, K. Y. Lee, and M. Misono, "Dehydration mechanism of ethanol in the pseudoliquid phase of H3-XCsXPW12O40," Journal of Molecular Catalysis A, vol. 55, pp. 293–301, 1989.

16. L. R. Pizzio, P. G. Vázquez, C. V. Cáceres et al., "C-alkylation reactions catalyzed by silica-supported Keggin heteropolyacids," Applied Catalysis A, vol. 287, no. 1, pp. 1–8, 2005.

17. H.-J. Kim, Y.-G. Shul, and H. Han, "Synthesis of heteropolyacid (H3PW12O40)/SiO2 nanoparticles and their catalytic properties," Applied Catalysis A, vol. 299, no. 1-2, pp. 46–51, 2006.

18. X. Zhao, Y. Han, X. Sun, and Y. Wang, "Structure and catalytic performance of H3PW12O40/SiO2 prepared by several methods," Chinese Journal of Catalysis, vol. 28, no. 1, pp. 91–96, 2007.

19. Q. Shu, J. Gao, Y. Liao, and J. Wang, "Reaction kinetics of biodiesel synthesis from waste oil using a carbon-based solid acid catalyst," Chinese Journal of Chemical Engineering, vol. 19, no. 1, pp. 163–168, 2011.

20. S. Saka and D. Kusdiana, "Biodiesel fuel from rapeseed oil as prepared in supercritical methanol," Fuel, vol. 80, no. 2, pp. 225–231, 1998.

21. D. Kusdiana and S. Saka, "Kinetics of transesterification in rapeseed oil to biodiesel fuel as treated in supercritical methanol," Fuel, vol. 80, no. 5, pp. 693–698, 2001.

22. K. Srilatha, R. Sree, B. L. A. Prabhavathi Devi, P. S. S. Prasad, R. B. N. Prasad, and N. Lingaiah, "Preparation of biodiesel from rice bran fatty acids catalyzed by heterogeneous cesium-exchanged 12-tungstophosphoric acids," Bioresource Technology, vol. 116, pp. 55–57, 2012.

23. L. Nakka, J. E. Molinari, and I. E. Wachs, "Surface and bulk aspects of mixed oxide catalytic nanoparticles: oxidation and dehydration of CH3OH by polyoxometallates," Journal of the American Chemical Society, vol. 131, no. 42, pp. 15544–15554, 2009.

24. J. C. Yori, J. M. Grau, V. M. Benítez, and J. Sepúlveda, "Hydroisomerization-cracking of n-octane on heteropolyacid H3PW12O40 supported on ZrO2, SiO2 and carbon Effect of Pt incorporation on catalyst performance," Applied Catalysis A, vol. 286, no. 1, pp. 71–78, 2005.

25. Q. Deng, W. Zhou, X. Li et al., "Microwave radiation solid-phase synthesis of phosphotungstate nanoparticle catalysts and photocatalytic degradation of formaldehyde," Journal of Molecular Catalysis A, vol. 262, no. 1-2, pp. 149–155, 2007.

26. S. A. Fernandes, A. L. Cardoso, and M. J. Da Silva, "A novel kinetic study of H3PW12O40: catalyzed oleic acid esterification with methanol via HNMR spectroscopy," Fuel Processing Technology, vol. 96, pp. 98–103, 2012.

AUTHOR NOTES

CHAPTER 2

Competing Interests

The authors declare that they have no conflict of interest in this publication.

Acknowledgements

This work was supported by the United States Department of Energy, Office of the Biomass Program under Contract No. DE-AC36-99GO10337 with the National Renewable Energy Laboratory (NREL).

CHAPTER 3

Acknowledgment

The authors gratefully acknowledge the financial support from the National Advanced Biofuels Consortium (NABC) and Department of Energy, USA (grant no. ZFT04064401). The authors would like to thank the Franceschi Microscopy and Imaging Center at Washington State University for SEM.

CHAPTER 4

Acknowledgments

This work was supported by the Bio4Energy research initiative (http://www.bio4energy.se), the Swedish Energy Agency (P35367-1), the Swedish Research Council (621-2011-4388), and the Biorefinery of the Future (http://www.bioraffinaderi.se).

Author Contributions

LJJ conceptualized, researched and wrote most of the manuscript. BA researched and wrote parts of the manuscript. NON conceptualized and critically revised the manuscript, and made most of the figures. All authors read and approved the final manuscript.

Competing Interests
LJJ and BA are co-authors on patent applications on detoxification.

CHAPTER 5

Competing Interests
The authors declare that they have no competing interests.

Author Contributions
MX and JY developed the idea for the study, and helped to revise the manuscript. HZ designed the research, did the literature review and prepared the manuscript. HZ, QL, and JY did experiments, plasmid construction, strain cultivation, Fed-Batch fermentation and product detection. All authors read and approved the final manuscript.

Acknowledgment
We greatly appreciate the financially support from the Qingdao Applied Basic Research Program (No. 12-1-4-9-(3)-jch), National Key Technology R&D Program (No. 2012BAD32B06), National High Technology Research and Development Program of China (863 Program, No. SS2013AA050703-2), National Natural Science Foundation of China (No. 21202179, 21376255), Knowledge Innovation Program of the Chinese Academy of Sciences (Y112131105).

CHAPTER 6

Acknowledgment
Financial support from Ministerio de Ciencia e Innovacion (Spain, Project MAT2010-21146) is gratefully acknowledged.

CHAPTER 7

Competing Interests
The authors confirm that this article content has no conflict of interest.

Acknowledgment

We thank Mr. Lucas Harrington for critical review of the manuscript. Funding for this review was supplied in part by the US Department of Energy under Contract DE-EE0003046 awarded to the National Alliance for Advanced Biofuels and Bioproducts (NAABB) and the Office of Energy Efficiency and Renewable Energy and Bioenergy Technologies (EERE-BETO). Funding was also provided by the Los Alamos National Laboratory Directed Research and Development Grant #LDRD20130091DR.

CHAPTER 9

Competing Interests

The authors have declared that no competing interests exist.

Author Contributions

Conceived and designed the experiments: BAR SPM. Performed the experiments: BAR SC MP DJB. Analyzed the data: BAR SC MP DJB SPM. Wrote the paper: BAR.

Acknowledgment

We thank Federico Unglaub and Susan Cohen for their technical and instructional support; and Susan and James Golden for the use of their microscope. We also thank Todd Peterson and Farzad Haerizadeh for thoughtful discussions.

CHAPTER 10

Competing Interests

The authors declare that they have no competing interests.

Author Contributions

CB designed and carried out this work, and drafted the manuscript. PS participated in experimental design, and molecular genetic studies and other experimental aspects of this work. JM participated in experimental design and strain construction. YY participated in genetic studies. SRC, HRB, and SWS supervised the research and edited the manuscript. NJH supervised the research, and wrote and edited the manuscript. All authors read and approved the final version of the manuscript.

Acknowledgment

This work was funded by the Department of Energy, Advanced Research Projects Agency-Energy (ARPA-E) Electrofuels Program, under contract DE-0000206-1577 to Lawrence Berkeley National Laboratory. This work was performed at the Joint BioEnergy Institute, which is funded by the Department of Energy, Office of Science, Office of Biological and Environmental Research under contract DE-AC02-5CH11231 to Lawrence Berkeley National Laboratory. We thank Vivek Mutalik for providing access to the Guava easyCyte Flow cytometry system. We thank Jonathan Vroom for assistance in constructing trfA mutations and the T7 stem-loop. We thank Yung Hsu Tang for assistance in constructing and testing pKT-Trfp, pCMTrfp, pCM271Trfp, and pCM271TcalRBSrfp.

CHAPTER 11

Acknowledgment

Dr Colmenares would like to thank the Marie Curie International Reintegration Grant within the 7th European Community Framework Programme and also the 2012–2014 science financial resources, granted for the international co-financed project implementation (Project Nr. 473/7. PR/2012, Ministry of Science and Higher Education of Poland). Rafael Luque gratefully acknowledges support from the Spanish MICINN via the concession of a RyC contract (ref. RYC 2009-04199) and funding under projects P10-FQM-6711 (Consejeria de Ciencia e Innovacion, Junta de Andalucia) and CTQ201-28954-C02-02 (MICINN).

CHAPTER 13

Acknowledgment

The authors are grateful to the Iran Nanotechnology Initiative Council (INIC) for its partial support of this project.

INDEX

Printed in the United States
by Baker & Taylor Publisher Services